Geography, History and Social Sciences

The GeoJournal Library

Volume 27

The titles published in this series are listed at the end of this volume.

Geography, History and Social Sciences

edited by

GEORGES B. BENKO

University of Paris I, Panthéon-Sorbonne, Department of Geography, Paris, France

and

ULF STROHMAYER

University of Wales, Department of Geography, Lampeter, U.K.
and
Fernand Braudel Center, State University of New York, Binghamton, NY, U.S.A.

KLUWER ACADEMIC PUBLISHERS
DORDRECHT / BOSTON / LONDON

A C.I.P. Catalogue record for this book is available from the Library of Congress

ISBN 0-7923-2543-5

Published by Kluwer Academic Publishers,
P.O. Box 17, 3300 AA Dordrecht, The Netherlands.

Kluwer Academic Publishers incorporates
the publishing programmes of
D. Reidel, Martinus Nijhoff, Dr W. Junk and MTP Press.

Sold and distributed in the U.S.A. and Canada
by Kluwer Academic Publishers,
101 Philip Drive, Norwell, MA 02061, U.S.A.

In all other countries, sold and distributed
by Kluwer Academic Publishers Group,
P.O. Box 322, 3300 AH Dordrecht, The Netherlands.

Cover
Le Penseur by Auguste Rodin,
exhibited at Musée Rodin, Paris.
Inventory number: S. 1295
photo by B. Jarret.

Printed on acid-free paper

Printed in the Netherlands

CONTENTS

Part IV: Economics

Part V: Politics

Part VI: Conclusion

PART I

INTRODUCTION

1. Geography, history and social sciences : An introduction

Georges Benko

«Societies are much messier
than our theories of them»
Michael Mann
The Sources of Social Power

1 Towards a unified social theory

Why are there communication problems between the different disciplines of the social sciences? And why should there be so much misunderstanding? Most probably because the encounter of several disciplines is in fact the encounter of several different histories, and therefore of several different cultures, each interpreting the other according to the code dictated by its own culture. Inevitably geographers view other disciplines through their own cultural filter, and even a benevolent view remains 'ethnocentric'. It was in order to avoid such ethnocentricity that Fernand Braudel called for more unity among the social sciences in 1958 : «I wish the social sciences...would stop discussing their respective differences so much...and instead look for common ground...on which to reach their first agreement. Personally I would call these ways : quantification, spatial awareness and 'longue duree'».

In its place at the center of the social sciences, geography reduces all social reality to its spatial dimensions. Unfortunately, as a discipline, it considers itself all too often to be in a world of its own. There is a need in France for a figure like Vidal de la Blanche who could refocus attention away from issues of time and space, towards space and social reality. Geographic research will only take a step forward once it learns to address the problems facing all the sciences. In the anglosaxon countries there is a new current emerging in periodicals such as *Society and Space* or *Antipode* which is attempting to do this with great

3

G. B. Benko and U. Strohmayer (eds.), Geography, History and Social Sciences, 3–12.
© 1995 *Kluwer Academic Publishers. Printed in the Netherlands.*

success. In France *Espaces et Societes, Espaces Temps* or *Geographie et Cultures* play the same role. The sociologist seeks refuge in words like environment and ecology to avoid saying geography, as they allow him to address the problems of space and what may be learned from its attentive observation. Spatial models reflect and to some extent explain social reality, and here especially the long term movements of all social categories. But astonishingly enough the majority of the social sciences disregard them. I have often thought that one of the things that the French social sciences had to boast about was the school of geography founded by Vidal de la Blanche, but sadly the spirit of his lessons has been betrayed. It is time for all the social sciences to find a place for the increasingly geographical concept of humanity which Vidal de la Blanche was already calling for in 1903.

Geography stands at the crossroads between the natural and the manmade sciences. Shakespeare said that all the world is a stage, but even if he was not referring to the world in its literal sense it is at least true that the world is a stage for human adventure. Geography is the study of this stage based on the premise that the stage affects human action. In this way geography is linked to all other social sciences. At the beginning of the century certain geographers claimed that geography played a preeminent and central role at the crossroads of the social sciences. However, like all imperial designs of this kind, it proved unsustainable and ridiculous and today the discipline is fragmented to the point that it might even be considered as a collection of hybrid sub-divisions. The division of geography into physical and human was inevitable, as was the splintering of the latter into sub-divisions. The systematic interaction with the social sciences has existed since the beginning of the 20th century. In the 1930's Bowman said that it was geography's synthetic knowledge of the region as a social reality which lead to its fraternisation with the historian, the economist and the sociologist. Indeed external factors have been the progressive force in geography. The introduction of the geographic environment to history is the proud achievement of the Ecole des Annales. Fernand Braudel and Lucien Lebvre were the most eminent theoreticians in geographical thinking. Braudel gives a central role to geography but he delves into economic history and anthropology, and into social geography.

Ideas flow more easily from the social sciences towards geography than in the opposite direction. Ideas from the realms of economics, history, anthropology and the social sciences have been important sources of inspiration for the progress of geography. The application of methods and techniques originating in the pure sciences (and already exploited by other disciplines within the social sciences) gave birth to what Peter Gould described as the *new geography* and the neo-positivist tendency. Conversely, the behaviourists have borrowed from psychology (perception, environment, living space, mental map etc..); in urban research geographers followed in the footsteps of sociologists and political scientists such as Burgess, Park, Castells, Lefebvre, Urry, Lash and Body-Gendrot. Geographers are on firm ground in the study of place, but their theoretical basis is provided exclusively by economists (Von Thünen, Weber, Lösch, Hotelling, Isard, Hoover, Lipietz and so on). The new generation of anglo saxon geographers is trying to integrate geography with the social sciences;

their points of reference are Foucault, Habermas, Derrida, Giddens and Lyotard.

The net result of all these tendencies is the incredible fragmentation which has caused human geography to encroach on the territory of other branches of the social sciences and physical geography to encroach on the natural sciences. As a discipline, geographers are turning increasingly towards the social sciences and are classified accordingly in the universities. It may be said these kinds of interaction with other scientific disciplines have maintained geography's momentum and saved it from inertia. As Dogan and Pahre (1991) have shown, most major innovations in the social sciences straddle the various disciplines. In fact the greater the innovation, the greater the chances of finding work marginal to one discipline but situated at the intersection with others. I propose that if geography exists, and is to maintain its identity, it may be found in a federal ideal, and not within the unity of one discipline; this federal ideal might be called social theory.

Although a historicist approach may suffice when trying to define the various branches of the social sciences, the same may not be said for the definition of social theory. The reason for this is simple : the concept of social theory is broader than sociology (and different to philosophy) and it disregards completely the institutional boundaries of the different disciplines. It is impossible to define the nature of social theory by referring to the meaning given to it by a particular author or by falling back on institutional practice. The classification of social theory is at present a theoretical impossibility. It follows that the history of social theory is theoretically informed and significant in two ways. First, it is the practice of social theory which determines the structural properties of history and the interpretation of the past. Second, the concept of social theory at work in written history serves as a model for the present definitions of social theory. Jeffrey Alexander (1982) classified the idea of a post-positivist vision of science in his presentation of what is perhaps its most advanced form : social theory as a collection of propositions covering a whole range of hypotheses, including non-empirical presuppositions. Furthermore Alexander stresses that the non-empirical dimension of social theory is composed of different aspects : methodology, models, ideologies, policies, moral beliefs and metaphysical and epistemological presuppositions. Since each aspect possesses it own character and has a unique impact on the formation of social theory, the exclusive identification of social theory with one of these aspects results in an unacceptable reductionist concept of social theory. All scientific fields are the result of a series of exercises, interpretation and synthesis. Knowledge is consolidated little by little, but the structure of knowledge is rarely perfectly coherent. In order to avoid the disciplinary fragmentation in the conceptualization of social reality, it is desirable to construct a global vision of the structure of society, and the quickest way of achieving this is by means of social theory.

There is nothing specifically new about this statement. Even at the turn of the century social theory was aware of what is now called the spatiality of social activity. Simmel was perhaps the first to propose a spatial form of sociology which deals with the way in which the fluctuating, fragmentary and contradictory social world of the modern metropolis is rooted in the volatile

circulation of money in time and space. The work of Durkheim on the division of
social work furthermore confers a central role to spatial structure and marks an
important stage in the development of sociology. However, the individual
interests of the age demanded that others respect the previous intellectual
divisions of the various fields of research. Rivalries in the universities have
created the compartmentalized fields of the social sciences rather than the
projections of capitalist social relations in space. Even topday, few transcend the
disciplinary squabbling, but I will mention two men who have, and shall present
their work in more detail.

Michel Foucault and Anthony Giddens are both representatives of the
modern social sciences. The striking achievement of Foucault's work is the
perfect integration of the historical contextualisation of human knowledge
through time. In the case of Giddens, it is the importance given to time and space
in a general social theory which is linked to the duality of action and structure.
These two cases will serve as illustrations of how a globalising social theory
approach can give relative satisfaction to researchers seeking a synthesis within
the social sciences.

2 The history of thought or the archaeology of knowledge ?

Traditionally the history of thought (in the social sciences) presupposes a
quasi continuity of knowledge. Michel Foucault's archaeology of thought
(1966) rejects the concept of continuity, considering instead knowledge to be a
means of knowing that was specific to each age and therefore evolutionary.
Such a shift of emphasis clearly opens up new areas for discussion.

The sub-title of the book clearly sets out its intention : to put together an
archaeology of the social sciences. The spirit of the initiative is to explain how
the emergence of the social sciences towards the beginning of the 19th century
fits into Western thinking. Michel Foucault demonstrates how the knowledge of
one age has its own specific codes which condition what may and may not be
said. Starting from the Renaissance, Foucault describes each period in history,
defining for each one the conditions governing the possibilities of knowledge
(the episteme), and in so doing he addresses the question : «Why is it that in a
given age, one could say one thing, while the other thing was never said ?»

The word archaeology refers specifically to knowledge which is
sandwiched between experience and science, and which 'determines the space
in which science and experience coincide and that in which they diverge'. For
example, it is within the domain of knowledge that we may explain and
understand how Quesnay and Condillac could both be products of the 18th
century. But one can no more hope to explain this in economic terms alone, than
one could by making sole reference to their methodology or subject of study.

In the passage about the relationship between material objects and words,
Foucault fuses the words/object dynamic, making it a fundamental feature of
knowledge and its corresponding episteme : this basic level which

simultaneously governs knowledge and the state of things known. By opposing practice with pure speculation one finds that they both rest on the same fundamental feature of knowledge. Since objects only exist in our knowledge in their empirical form, preceding theoretical discourse, and since the theories of one age are linked to the practices of the same age, an interpretation which leaps from practice to theoretical discourse is insufficient. There is a more general system which is the fundamental necessity of knowledge that needs to be invested both in theory and in practice. Hence the question no longer relates to the relationship between facts, material events, techniques and their theorisation, but to the episteme of knowledge itself and its relationship to the practices and theory which it engenders.

One might establish a parallel between the positions of Foucault and those of T. S. Kuhn (1962). Both are relativists, both reject the idea of an immutable truth, waiting silently to be unraveled bit by bit by hesitant, clumsy, stammering but tenacious academics. However, the comparison ends here. The Kuhnian paradigm applies to the so-called natural sciences while the archaeology refers to the social sciences. «The Structure of Scientific Revolution» focuses its analysis on the degree of acceptance of normal science at any given moment and finally rests upon a sociology of the scientific community. The archaeology on the other hand, refers to the very act of knowing. Foucault seeks to explain knowledge, not by sociology, i.e. by something which is different to it, but rather by comparing like with like and explaining knowledge by its episteme. As the term archaeology suggests he is working at a profounder and more hidden level. That is not to say he is returning to the *a priori* categories of Kantian rationalism. The archaeological analysis certainly tries to show *a priori* categories at work in Western culture, but they are historically dated and defined *a posteriori* by the positivism they represent.

3 Space, time and structuration

In order to grasp the structuralist concept, we must not neglect the concepts of time and space as analyzed by Anthony Giddens for he engendered a marriage between history and geography that was engineered by a social scientist.

The analysis of time and space plays a central role in structuration theory in which structural concepts are linked to the duality of action and structure. Social systems, in other words, are not just structured by rules and resources, but they also occupy space and time. The structural properties of social systems only exist if forms of social behaviour repeatedly reproduce themselves in time and space. All human actions are bound through time to a time/space structure. This may sound banal, but as Giddens explains, in the past social scientists neglected not only the temporality of social behaviour but also its spatial distribution; these notions were considered more as the environments in which social actions took place.

The concepts of space and time appeared for the first time in 1979 in *Central Problems in Social Theory* but they became social themes in Giddens's later work. Giddens wanted to show how space and time are absolutely fundamental to the study of social habits. At the same time this was an attempt to break down the conventional academic boundaries between sociology, geography and history. In his theory of structuration, Giddens considers the time/space dynamic as the building blocks of social systems. This dynamic must not be considered as the context for action, but as an expression of the nature of action. Giddens underlines the contextual nature of action : the fact that human action, in exactly the same way as learning, lasts for a certain amount of time during which individuals evolve and rationalize their actions. All dimensions of the contexts for action take place routinely in space and time.

Certainly Giddens was fascinated by time which he considered to be the most enigmatic element of human existence. He was strongly attracted by Heidegger's work on time, but the most decisive influence came from the time/space geography developed by Hägerstrand. Hägerstrand's approach is based on the identification of the sources of constraints on human activity (coupling constraints and packing constraints) which are linked to the nature of the body and to the physical contexts in which these activities are carried out. Hägerstrand's time/space geography takes its cue from the routine nature of daily life. Routines are the expression of the repetitiveness of daily life. The majority of interactive elements are rooted in time, and one can only grasp their significance by taking into account their routine and repetitive nature. Day-to-day life has a set duration; repetition passes the time. Social actors are situated in time/space, moving through their lives along what Hägerstrand calls their spatio-temporal paths. The dual structure presupposes the reflexive control of these social actors throughout their daily social interaction. Routine is the founding concept of structuration theory. Routine is intrinsic both to the maintenance of the personality of the social actor as he proceeds along the path of his daily activities, and to the social institutions, the institutional character of which depends entirely on their continuous output. The typical displacement models of each person may be considered as the repetitive, routine activities undertaken on a daily basis. Individuals interact with one another. Simultaneously they move within physical contexts whose characteristics interact with the capacities of each individual. According to Giddens the continuity of daily life depends largely on routine interactions between people who co-habit in time and space. Meetings play a central role in the interactions. They bring into play spatialisation; first in relation to the position of the body, and then the sequential spatialisation of each contribution to a meeting viewed sequentially, in the order of play. The order of play in meetings is judged by Giddens as fundamental to the study of interaction; it expresses the essential dimensions of the nature of the interaction, and furthermore, as the dominant feature of the sequential nature of social activity, it reflects the universal nature of social reproduction. Social development also includes spatio-temporal movement : the most significant present day example being the global expansion of western industrial capitalism.

Despite certain criticisms of Gidden's work, the theory of structuration is not eclectic but rather a theoretical synthesis, where eclecticism involves the juxtaposition of mechanical elements and theoretical synthesis combines the same elements in such a way that their combination produces a new fusion which is qualitatively distinct from each of its component parts. This new combination which contains hypotheses and concepts as well as its own principles constitutes a new basis for research. Finally, we may add that Giddens considers all his work as part of an ongoing project : «the making of structuration theory».

4 Conclusions

The relationships between space and the social sciences are complex and researchers from various disciplines seek to identify, evaluate and explain them. In these turbulent, yet progressive times for the social sciences, space is playing an increasingly important role. This book not only reaffirms the concept of space in the social sciences, and its pivotal role in linking different disciplines, but it also identifies geography's origins within the social sciences.

Following the epistemological revival, the challenge facing geography relates to its theoretical base. Will geographers be capable of outgrowing their old empirical ways and their impoverished generalist approach, and of constructing systems capable of taking the unique into account ? In order to move in this direction, we can rely on certain principles : a dimensional approach to the problematics of the social sciences, in which each one is like a transversal section of social reality; a statute for dealing with unique processes -as exists in all sciences - which should not be considered as exceptions to the rule, but as products of causal systems which are sufficiently complex to encompass them; an interest in confronting different metric and spatial systems, leading to the construction of a universal social space which one may compare with others obtained from totally different starting points.

Geography is asserting itself as a social science, so it must make sure that it keeps an ear open to any messages capable of stimulating its theoretical imagination, but she must also take care not to lose the hard-earned equality with its neighbours. This work will, we hope, contribute to the dialogue taking place within the social sciences, and more specifically between geography and history - or, to put it more simply, to the dialogue between space and time.

The approaches of the various authors presented in this volume are very diverse. We have been able to group the contributions into four principal sections. The section entitled 'Geographical Evolutions' contains five completely different views of the ways in which geography and the concept of space may be linked with history. Firstly much depends on the way in which society and social theory is conceived. In the 1960s Michel Foucault confronted the 19th century obsession with time with the shift of interest in the 20th century towards space. He reminds us that we are living in an age in which proximity and distance, and cohesion and dispersal are simultaneous. In his opinion our experience of the world is less that of a long life unfolding through time, but of a network linked to a complex juncture of points and interactions. Derek Gregory bases his analysis on the work of Henri Lefebvre (who is enjoying a revival of his popularity in the United States after the belated translation of *The Production of Space*). He refers to Lyotard and conducts a critical debate between the various authors. Martin Hampl - one of the rare exponents of theoretical geography in Central Europe - studies geographical systems and attempts an abstract synthesis of geography into a mathematical language. The space opened between these two latter contributions, we hope, will come to mark the breadth of possibilities of a new and theoretically informed geography. Jean-François Staszak and Milton Santos also stand at two temporarily opposite poles : their contributions signify the «beginning» and the «end» of geography. One part, hellenic thought - a return to the origins of human thought - brings us to the starting point of the representation of the world, while the other finds us in the world of hypermodernity in global space and time.

The second major section is devoted essentially to a historical - and essentially cultural - view of cities. Paul Claval demonstrates the influence of the urban environment on the evolution of artistic creativity through the history of Parisian life at the beginning of this century. Clyde Weaver presents the history of the controversial relationship between town and country in a truly potted history of ideas, with examples drawn from several centuries. Torsten Hägerstrand, as major scholar, builds his article around the key concepts of landscape geography by evoking its use through history (its rise and fall) «its traps» and its interest today. Ulf Strohmayer illustrates the spatial construction of modernity in Germany at the beginning of the century and its ideological utilization.

The two final sections relate to economics and politics. The Regulation School is situated within the broad spectrum of historical economics. The approach adopted has allowed us to provide a coherent and coordinated collection of analyses of the dynamics of yesterday's and today's capitalism. Mick Dunford and Diane Perrons have taken their inspiration from the regulationists and applied it brilliantly to regional development. Georges Benko reviews the recent evolutionary change in economic theory and Milton Santos writes of the globalisation and fragmentation of our contemporary world. In the spirit of Jacques Levy, politics must be apprehended in its different processes : representation, legitimisation, action; it is only by articulating their different spatialities that one may arrive at an image of political space. Ron Johnston develops two concepts which are essential to political geography : territory and State, and he analyses their complex relationship. Judith Lazar runs through the principal theories linking space and time in the social context.

«Geography must be at the heart of all I do», said Foucault in 1976. But for this statement to become a credible one for a number of researchers in the social sciences, we must overcome the specifically modern impasse in social theory. In order to do so we must build a human geography which is sensitive both to the differences and the distances which are buried in the fragile constitution of social reality.

Note :

My special thanks to Mick Dunford for his invaluable help in the production of the English version, and to Melanie Knights for its translation.

References

Agnew, J., 1990, 'Les lieux contre la sociologie politique', *Espaces Temps,* 43/44, 87-94
Alexander, J., 1982, *Theoretical Logic in Sociology,* Los Angeles : University of California Press
Auriac, F. and Brunet, R., eds., 1986, *Espaces, jeux et enjeux,* Paris : Fayard
Benko, G. B., ed., 1988, *Les nouveaux aspects de la théorie sociale. De la géographie à la sociologie,* Caen : Paradigme
Bourdieu, P., 1984, 'Réponse aux économistes,' *Economies et Sociétés,* 18, 10, 23-32
Boyer, R., 1989, 'Economie et histoire : vers de nouvelles alliances', *Annales ESC,* 44, 6, 1397-1426
Braudel, F., 1958, 'Histoire et sciences sociales. La longue durée', *Annales ESC,* 4, 725-753
Braudel, F., 1969, *Ecrits sur l'histoire,* Paris : Flammarion
Bryant ,C. G. A. and Jary, D., eds., 1991, *Giddens' theory of structuration : A critical appreciation,* London, Routledge
Bunkse, E., 1990, 'Saint-Exupéry's Geography Lesson : Arts and Science in the Creation and Cultivation of Landscape Values', *Annals of the Association of American Geographers,* 80, 1, 96-108
Buttimer, A., 1990, 'Geography, Humanism, and Global Concern', *Annals of the Association of American Geographers,* 80, 1, 1-33
Buttimer, A., 1993, *Geography and the human spirit,* Baltimore : John Hopkins University Press
Claval, P., 1980a, *Les mythes fondateurs des sciences sociales,* Paris : PUF
Claval, P., 1980b, 'Epistemology and the history of geographical thought', *Progress in Human Geography,* 4, 3, 371-384
Claval, P., 1993, *La géographie au temps de la chute des murs,* Paris : L'Harmattan
Cox, K. R., 1990, 'Classes, localisation et territoire', *Espaces Temps,* 43/44, 95-102
Dogan, M. and Pahre, R., 1991, *L'innovation dans les sciences sociales. La marginalité créatrice,* Paris, PUF
Dose, F., 1985, 'Foucault face à l'histoire', *Espaces Temps,* 30, 4-22
Driver, F., 1985, 'Power, space, and the body : a critical assessment of Foucault's 'Discipline and Punish'', *Environment and Planning D : Society and Space,* 3, 4, 425-446
Foucault, M., 1966, *Les mots et les choses. Une archéologie des sciences humaines,* Paris, Gallimard
Foucault, M., 1969, *L'Archéologie du savoir,* Paris: Gallimard
Foucault, M., 1976, 'Questions à Michel Foucault sur la géographie', *Hérodote,* 1, 71-85
Giddens, A., 1979, *Central Problems in Social Theory,* London : Macmillan
Giddens, A., 1984, *The Constitution of Society,* Cambridge : Polity Press
Giddens, A., 1987, *Social Theory and Modern Sociology,* Stanford : Stanford University Press
Gregory, D., 1994, *Geographical Imaginations,* London : Blackwell
Harvey, D., 1990, 'Between Space and Time : Reflections on the Geographical Imagination', *Annals of the Association of American Geographers,* 80, 3, 418-434
Hayek, F. A., 1956, 'The Dilemma of Specialization', in White, L. D., ed., *The State of Social Sciences,* Chicago : University of Chicago Press, 462-473
Held, D. and Thompson J. B., eds., 1989, *Social Theory of Modern Society,* Cambridge : Cambridge University Press

Hippel, E. von, 1988, *The Sources of Innovation,* Oxford : Oxford University Press

Kirby, A. M., 1986, 'Le monde braudellien', *Environment and Planning D : Society and Space,* 4, 2, 211-219

Knight, D. B., 1982, 'Identity and Territory : Geographical Perspectives on Nationalism and Regionalism', *Annals of the Association of American Geographers,* 72, 2, 514-531

Kuhn, T. S., 1962, *The Structure of Scientific Revolution,* Chicago : The University of Chicago Press

Kuhn, T. S., 1979, 'The Relations Between History and History of Science', in Rabinow P. and Sullivan W. M., eds., *Interpretive Social Science. A Reader,* Los Angeles : University of California Press, 267-300

Lazar, J., 1992, 'La compétence des acteurs dans la théorie de la structuration de Giddens', *Cahiers Internationaux de Sociologie,* 2, 399-416

Mann, M., 1986, *The Sources of Social Power,* vol 1 : *A History of Power from the Beginning to A.D. 1760,* Cambridge : Cambridge University Press

Noiriel, G., 1989, 'Pour une approche subjectiviste du social', *Annales ESC,* 44, 6, 1435-1460

Roncayolo, M., 1989, 'Histoire et géographie : les fondements d'une complémentarité', *Annales ESC,* 44, 6, 1427-1434

Sack, R., 1980, 'Conceptions of geographic space', *Progress in Human Geography,* 4, 3, 313-345

Sack, R., 1983, 'Human Territoriality : A Theory,' *Annals of the Association of American Geographers,* 73, 1, 55-74

Sack, R., 1986, *Human territorality : its theory and concepts,* Cambridge : Cambridge University Press

Seidman, S., 1983, *Liberalism and the origins of European Social Theory,* Berkeley, CA : University of California Press

Shotter, J., 1985, 'Accounting for place and space', *Environment and Planning D : Society and Space,* 3, 4, 447-460

Storper, M., 1985, 'The spatial and temporal constitution of social action : a critical reading of Giddens', *Environment and Planning D : Society and Space,* 3, 4, 407-424

Stohmayer, U., 1993a, 'Beyond theory : the cumbersome materiality of shock', *Environment and Planning D : Society and Space,* 11, 3, 323-347

Strohmayer, U., 1993b, 'Modernité, post-modernité ou comment justifier un savoir géographique', *Géographie et Cultures,* 6, 75-84

Tuan, Y-F., 1990, 'Realism and Fantasy in Art, History and Geography', *Annals of the Association of American Geographers,* 80, 3, 435-446

Vidal de La Blache, P., 1903, 'La géographie humaine, ses rapports avec la géographie de la vie' *Revue de synthèse historique,* 7, 219-240

Wallerstein, I., 1985, 'Vers une recomposition des sciences sociales', *Espaces Temps,* 29, 36-42

Wallerstein, I., 1988, 'Faut-il 'dé-penser' les sciences sociales du XIXe siècle', *Revue Internationale des Sciences Sociales,* 118, 579-585

Ward, D., 1990, 'Social Reform, Social Surveys, and the Discovery of the Modern City', *Annals of the Association of American Geographers,* 80, 4, 491-503

White, L. D., ed., 1956, *The State of Social Sciences,* Chicago : University of Chicago Press

Williams, C. and Smith, A. D., 1983, 'The national construction of social space', *Progress in Human Geography,* 7, 4, 502-518

Georges Benko
Université de Paris I – Panthéon-Sorbonne
191, rue Saint-Jaques
75005 Paris
France

PART II

SPATIAL THINKING IN HISTORY

2. Lefebvre, Lacan and the production of space

Derek Gregory

> «In advance of the invasion we will hear from the depths of mirrors the clatter of weapons.»
>
> Jorge Luis Borges
> *The Book of Imaginary Beings*

Introduction

In this essay I offer a partial, necessarily provisional reading of Henri Lefebvre's account of the production of space through the *grille* of concepts provided by Lacanian pschoanalytic theory. This is a dificult undertaking but also, I imagine, an unusual and unexpected one, and so I need to spell out my reasons for doing so as carefully as I can.

Lefebvre's writings represent (for me) one of the most important single sources for the development of a critical understanding of the production of space. In the course of his long life he wrote almost seventy books, but it was those that appeared between 1968 and 1974 that addressed most directly what came to be called «the urban question». *La production de l'espace,* which was first published when Lefebvre was in his early seventies, is the climax of that sequence and reveals him at the height of his powers : imaginative, incisive and immensely suggestive.[1] His interventions have to be situated in time and space, of course, and their grounding in the events of May 68 in France is of the first importance : I do not mean to imply that his claims can be extrapolated to other

---

[1] Henri Lefebvre, *La production de l'espace* (Paris: Anthropos, 1974), translated into English by Donald Nicholson-Smith as *The production of space* (Oxford, England and Cambridge, Mass: Blackwell Publishers, 1991).

G. B. Benko and U. Strohmayer (eds.), Geography, History and Social Sciences, 15–44.

settings without critical scrutiny or careful reconstruction. Even so, the contemporary interest in the spatiality of social life and the reassertion of space in critical social theory have together turned Lefebvre into a pivotal figure in debates on both sides of the Atlantic.

And yet the Anglophone reception of his work has long been problematic. Most British and American commentators seem puzzled by it. In Perry Anderson's *Considerations on Western Marxism,* for example, Lefebvre is portrayed as the lonely philosopher. Read by the exiled Benjamin in Paris, he was himself «an international isolate at the close of the thirties; within France itself, his example was a solitary one.» In the course of his second voyage *In the tracks of historical materialism,* Anderson discovers a topographical break, a shift in the locus of politico-intellectual production from continental Europe to what he sees as traditionally the most backward zones of Marxism's cultural geography, Britain and North America; there he charts «the rise of historiography to its long overdue salience within the landscape of socialist thought as a whole», the emergence of an «historically centred Marxist culture»; and yet he still records Lefebvre, like some Ancient Mariner, as «the oldest living survivor» of his previous voyage, «continuing to produce imperturbable and original work on subjects typically ignored by much of the Left.» A note reveals that Anderson has in mind Lefebvre's *La production de l'espace :* but apart from this bibliographical marker-buoy, submerged at the bottom of the page, those subjects are ignored by Anderson too.[1]

But there are one or two exceptions. The most important for my present purposes is Fredric Jameson who, to continue the marine metaphor, mercifully does not see a continued interest in Marxism as a ticket on the *Titanic.* He has repeatedly applauded Lefebvre's work, and in fact once complained in an acerbic aside that Lefebvre's «conception of space as the fundamental category of politics and of the dialectic itself - the one great prophetic vision of these last years of discouragement and renunciation - has yet to be grasped in all its pathbreaking implications.» He regards the English translation of *La production de l'espace* as an event which «might well play a decisive role in reorienting the problems and debates of contemporary philosophy.»[2] Jameson is not the only critic to have used Lefebvre to underwrite the «predominance of space in the postcontemporary era», a claim which continues to trouble me, but I focus on his intervention here because it is particularly sensitive to the radical implications of Lefebvre's project. «In effect,» Jameson remarks,

> Lefebvre called for a new kind of spatial imagination capable of confronting the past in a new way and reading its less tangible secrets off the template of its spatial structures - body, cosmos, city, as all those marked the more intangible organization of cultural and libidinal

---

[1] Perry Anderson, *Considerations on Western Marxism* (London: Verso, 1979) p. 37; Perry Anderson, *In the tracks of historical materialism* (London: Verso, 1983) pp. 24, 30.

[2] Fredric Jameson, 'Architecture and the critique of ideology', in Joan Ockman, Deborah Berke and Mary Mcleod (eds), *Architecture, criticism, ideology* (Princeton: Princeton Architectural Press, 1985) pp. 51-87; the quotation is from p. 53. His endorsement of *Production* appears on the dust-jacket of the English translation.

economies and linguistic forms. The proposal demands an imagination of radical difference, the projection of our own spatial organizations into the well-nigh science-fictional and exotic forms of alien modes of production.[1]

I should say at once that I do not introduce Jameson as a way of claiming Lefebvre for postmodernism and its supposed «supplement» of spatiality - on the contrary, I think there are real problems in trying to do so - but because Jameson's work also draws on Lacanian psychoanalytic theory (which is why this passage speaks of «libidinal economies and linguistic forms»), and it is precisely the tense conjunction of Lefebvre and Lacan that is my primary concern in this essay.

Jameson's interest in Lacan is double-edged. In his early writings it is both direct and indirect, mediated by Althusser's appropriations of Lacan that play a strategic part in his formulation of *The political unconscious.*[2] In Jameson's later writings, and particularly those that address the production of space within the late twentieth-century culture of capitalism, Lacan becomes much more important in his own right : his ideas are used to explore the postmodern dislocation of the signifying chain, to illuminate the contemporary disjunction between body and space, and to suggest the need for an aesthetic of «cognitive mapping» that can provide an intrinsically spatial model of political culture.[3] But when Jameson suggests in the same essay that, for this newer, presumably postmodern aesthetic in the visual arts, the representation of space has come to be felt as incompatible with the representation of the body», he is, I think, drawing on both Lefebvre and Lacan.[4]

The reason for my interest in this double articulation is that in *La production de l'espace* Lefebvre sketches two, closely imbricated «histories of space». The most prominent is a history of the social relations between human bodies and spaces which is constructed through a radicalisation of Marx's critique of *political economy.* Most commentators fix Lefebvre's intellectual

---

[1] Fredric Jameson, *Postmodernism or the cultural logic of late capitalism* (Durham: Duke University Press, 1991) pp. 364-5. I intend to consider the purchase of Lefebvre's ideas on late twentieth-century spatialities in another essay, where I will pay particular attention to the spaces of representation opened up by contemporary science fiction. For an intermapping of science fiction and urban critique - and of bodies and (cyber)spaces - which draws suggestively on Lefebvre, see Scott Bukatman, *Terminal identity: the virtual subject in post-modern science fiction* (Durham: Duke University Press, 1993).

[2] Scott Bukatman, *The political unconscious* (Ithaca: Cornell University Press, 1981); see also Scott Bukatman, 'Imaginary and Symbolic in Lacan: Marxism, psychoanalytic criticism and the problem of the subject', *Yale French Studies* 55/6 (1977) pp. 338-395. For an introduction to *The political unconscious,* see William Dowling, *Jameson, Althusser, Marx* (Ithaca: Cornell University Press, 1984).

[3] See in particular sections III and VI of the title essay in Jameson, *Postmodernism* (1991) 'Cognitive mapping' is a troubled term. Jameson takes it from Kevin Lynch's classic *The image of the city* (Cambridge: MIT Press, 1960) - which is a long way from either Althusser or Lacan - and one critic suggests that Jameson's misadventures constitute 'a case of the analyst losing his place amidst the scenographies generated by his analysand': Donald Preziosi, 'La vi(ll)e en rose: Reading Jameson mapping space', *Strategies* 1 (1988) pp. 82-99; the quotation is from p. 83.

[4] Preziosi (1988) p. 34.

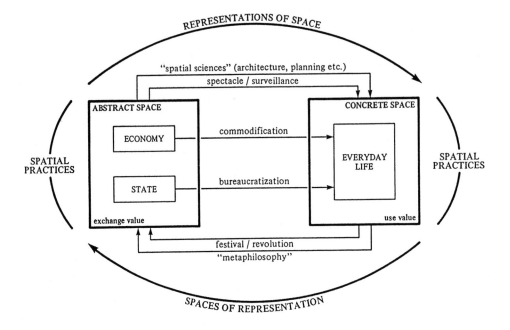

Figure 2.1: The colonisation of everyday life (Gregory, 1994)

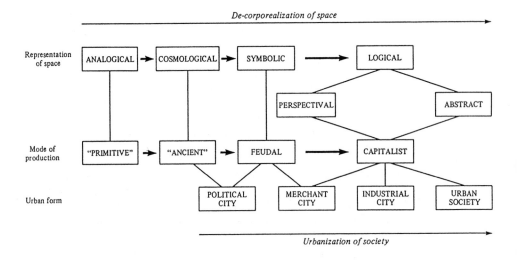

Figure 2.2: The decorporealization of space (Gregory, 1994)

topography by sighting these co-ordinates, plotting the lines between his Marxism and an eclectic humanism, and I do not mean to diminish their importance. Lefebvre's position within Western Marxism may be as singular as Anderson says, but there can be no doubt of his commitment to a reconstructed historical materialism. Indeed, there are suggestive (though I think superficial) parallels between Habermas's theory of communicative action, which he once advertised as a «reconstruction» of historical materialism, and Lefebvre's account of the production of space. Habermas speaks of the colonisation of the life-world by the system, Lefebvre of the colonisation of the spaces of «everyday life» by the production of an abstract space (Figure 1). But Habermas sees this process as a deformation of the project of modernity, which his own work is intended to reclaim, and his argumentation sketches and reconstructive histories are designed to expose the inner logics and circuit diagrams of a transcendent process of rationalization.[1] In contrast, Lefebvre strongly implies that the triumph of functionalist reason is inscribed within the very heart of capitalist modernity. Its impositions and alienations are not purely conjunctural or contingent, and Lefebvre offers an alternative history whose progressions install a dramatically different historicism (Figure 2). From this perspective the colonisation of everyday life has been advanced through a «logic of visualization» in which the spatialities of a fully human existence have been almost entirely erased : all that remains are traces, feint human geographies in an otherwise abstract geometry. If capitalism is constituted as a particular constellation of power, knowledge and spatiality, as Lefebvre seems to say, then its modernity is inscribed in a decorporealized space.[2]

    Although Lefebvre is concerned to chart the historical succession and superimposition of modes of production of space, his project is not a simple extension of Marx's critique of political economy. He remains indebted to Marx's writings, and most particularly to the sketches contained in the *Grundrisse,* but he believes that there is something sufficiently distinctive about the collective production of space under neo-capitalism that makes it necessary to transcend the classical categories of historical materialism. Partly for this reason there is, I think, another history of space in Lefebvre's work, which is derived from the developmental relations between the human body and space - from those phases through which the human being is supposed to move during its passage from infancy to adulthood - and which is constructed through an oblique, and at times almost subterranean critique of Lacanian *psychoanalysis.* In my previous discussions of Lefebvre, like most commentators I paid relatively little attention to his response to Lacan. But it is impossible to read the same book twice, and a re-reading of *La production de l'espace* has since persuaded me of the importance of this psychoanalytic provocation. In what follows, therefore, I want to show how Lefebvre's reworking of historical materialism, and in

---

[1] Cf. Jürgen Habermas, *The theory of communicative action.* Vol. 2: *The critique of functionalist reason* (Cambridge: Polity Press, 1987); this was first published in German in 1981.

[2] For a fuller discussion, see 'Modernity and the production of space', in my *Geographical imaginations* (Oxford, England and Cambridge, Mass: Blackwell, 1994) pp. 348-416.

particular his construction of concepts of spatiality, depends in crucial ways on these unstable foundations.

Lefebvre and Lacan

In many ways, I realise, parallels between the two projects are unremarkable. On the one side, Popper castigated both pschoanalysis and historical materialism as pretenders to the throne of Science, as «pseudo-sciences» whose propositions scrupulously avoided his own criteria of falsification.[1] On the other side, modern French thought has often been preoccupied with conjoining Freud and Marx, libidinal economy and political economy, and there is a vital tradition of what Margaret Cohen has recently described as «Gothic Marxism» that from the early twentieth century sought to appropriate Freud's writings for a renewed historical materialism.[2] According to Mark Poster, Lefebvre was probably the first French philosopher to read Freud seriously, but he claims that he did so «only during his brief interest in surrealism in the 1920s.»[3] Certainly, the surrealists were the first group in France to draw on Freud's work in any sustained and constructive fashion - one writer suggests that Freud was almost unknown in France until the surrealists «discovered» him, though whether this did very much to further Freud's reception is an open question - and there were plainly important connections between psychoanalysis and Breton's so-called «modern materialism».[4] But Lefebvre's (critical) interest in surrealism was not eclipsed in the inter-war period; it is vividly present in his writings after 1958 (when he broke with the PCF, the French Communist Party) and in *Production,* written in the wake of May '68, he concedes that «surrealism appears quite otherwise than it did half a century ago»

---

[1] Karl Popper, *The logic of scientific discovery* (London: Hutchinson, 1959); this was first published in German in 1934-5.

[2] Margaret Cohen, *Profane Illumination: Walter Benjamin and the Paris of surrealist revolution* (Berkeley: University of California Press, 1993). Closer to our own time, Vincent Descombes has suggested a reorientation in French critical thought, post 68, in which historical materialism was supposed to be revivified 'with an injection of desire and *jouissance.'* He notes the (passing) importance of Herbert Marcuse and hence, indirectly, of the Frankfurt School of critical theory, but pays much more attention to the contributions of Gilles Deleuze and Jean-François Lyotard. See Vincent Descombes, *Modern French Philosophy* (Cambridge: Cambridge University Press, 1980) pp. 171-186.

[3] Mark Poster, *Existential Marxism in post-war France: from Sartre to Althusser* (Princeton: Princeton University Press, 1975) p. 260.

[4] Helena Lewis, *The politics of surrealism* (New York: Paragon, 1988) p. x. Freud's response to his 'discovery' was equivocal. 'I am not able to clarify for myself what surrealism is and what it wants,' he wrote to Breton, and Cohen suggests that he was disconcerted by Breton's interest in 'statements by Freud that pivot suggestively to Marxist theory.' See Cohen, Profane illumination (1993) pp. 57-61 and, for a further discussion, Elizabeth Roudinesco, 'Surrealism in the service of psychoanalysis', in her *Jacques Lacan & Co. A history of psychoanalysis in France, 1925-1985* (Chicago: University of Chicago Press, 1990) pp. 3-34.

and that its theoretical promise, although unrealized, remains none the less vital : «to decode inner space and illuminate the nature of the transition from this subjective space to the material realm of the body and the outside world, and thence to social life.»[1] As I will show, this was, in essence, the object of Lefebvre's own «spatial architectonics».

But neither was Lefebvre's interest in psychoanalysis confined to Freud, and throughout *Production* he conducts a running skirmish against Lacan's provocative re-reading of Freud. Lefebvre and Lacan were exact contemporaries, both born in 1901. Like Lefebvre, Lacan moved in the circles - sometimes the shadows - of surrealism, and in fact the discussion between Bataille, Caillois and Leiris that led to the foundation of the Collège de Sociologie, a group of dissident surrealists, took place in 1937 in Lacan's Paris apartment. He was drawn to surrealism by its obsessive play with language; as David Macey remarks, surrealism involved an exploration of the production of signification - of what Lacan would later call the symbolic - and was thus, in part, a challenge to conventional conceptions of subjectivity.[2] The surrealists were also among the first in France to reclaim Hegel for revolutionary thought, and Lefebvre's involvement with surrealism thus led him not only to Freud, as I have noted, but also to Hegel (and then through him to Marx). Lefebvre's was no simple Hegelian Marxism, and his appropriation was a stubbornly critical one, but the traces of Hegel are indelibly present throughout his work and are particularly prominent in his account of the production of space.[3] The Collège de Sociologie played a part in introducing Lacan to Hegel, whose ideas proved to be a vital spur to the development of his own thesis about the constitution of subjectivity. But both Lefebvre and Lacan, like many others of their generation, were also profoundly influenced by Kojève's recuperation of Hegel's *Phenomenology,* and Macey suggests that if one were to develop a cultural geography of dissident Paris in the 1930's, then «Kojève's lecture room would lie at its heart.»[4] Lefebvre and Lacan learned different lessons there, to be sure, but

---

[1] Lefebvre, *Production* (1991) p. 18.

[2] For an elaboration of the (complex) connections between surrealism and Lacan's own work, see Cohen, *Profane illumination* (1993), pp. 147-153 and David Macey, 'Baltimore in the early morning', in his *Lacan in contexts* (London: Verso, 1988) pp. 44-74. The title of Macey's essay is taken from what he regards as Lacan's 'surrealist' image of the unconscious as 'Baltimore in the early morning'.

[3] Soja argues that 'Hegel and Hegelianism promulgated a powerful spatialist ontology' that was lost when Marx subsequently 'inverted' Hegel. 'The early expansion of Marxism in France, however, coincided with a major Hegelian revival, a reinvestiture that carried with it a less expurgated sensitivity to the spatiality of social life.' Its leading bearer, so he claims, was Lefebvre. See Edward Soja, *Postmodern geographies: the reassertion of space in critical social theory* (London: Verso, 1989) pp. 46-7, 86. For a fuller discussion of the filiations between surrealism, Hegel and Lefebvre - though one which says nothing about spatiality - see Martin Jay, 'Henri Lefebvre, the surrealists and the reception of Hegelian Marxism in France', in his *Marxism and totality: adventures of a concept from Lukács to Habermas* (Cambridge: Polity Press, 1984) pp. 276-299.

[4] Macey, *Lacan* (1988), pp. 96. For discussions of Kojève's role in the Hegelian revival in France, see Judith Butler, *Subjects of desire: Hegelian reflections in twentieth-century France* (New York: Columbia University Press, 1987) and Michael Roth, *Knowing and history: appropriations of Hegel in twentieth-century France* (Ithaca: Cornell University Press, 1988).

where they differed most decisively was over historical materialism. Lacan's earliest political sympathies were with the Right rather than the Left, and when he was subsequently drawn into a closer association with Marxism it was with an avowedly structural Marxism that Lefebvre strenuously repudiated. In fact, it was the principal architect of structural Marxism, Louis Althusser, who invited Lacan to move his seminar to the École Normale Supérieure in 1963, and Althusser later used Lacan's work to construct a theory of ideology that was at odds with Lefebvre's humanist inclinations.[1] Many critics have claimed that it was also at odds with Lacan - that no matter how closely Althusser read *Capital,* he did not accord Lacan the same careful study[2] - but it was still the case that by 1966 Lacan was endorsing Althusser against Sartre, at a time when Lefebvre and Sartre had already established an intellectual rapprochement of sorts.[3] By the time the events of May 68 were unfolding on the streets of Paris Lacan and Lefebvre were on the same side, more or less, and for that matter on the same sidelines, but their positions were none the less different. Both were accused of fanning the flames of student unrest, though neither of them joined the students on the barricades, but Lefebvre was much more sympathetic and his classes at Nanterre were a rallying-ground for many of the young revolutionaries, whereas Lacan stormed out of a lecture at Vincennes denouncing those who longed for «a Master».[4]

These biographical connections and affiliations are important, but none of them disclose the conceptual tensions between Lefebvre and Lacan in any detail. For this reason I want to set out in summary form a series of claims made by Lacan (at various times) from which Lefebvre most conspicuously dissents in *La production de l'espace.* I hope it will be obvious that these spare observations are not - cannot be - a synoptic account of Lacan's work : that would be absurd. They do not constitute a critique of Lacan either, but my presentation of these ideas should not be confused with their endorsement. By

---

[1] I have found Martin Jay particularly helpful on the relations between Lacan and Althusser: see 'Lacan, Althusser and the specular subject of ideology', in his *Downcast eyes: the denigration of vision in twentieth-century French thought* (Berkeley: University of California Press, 1993) pp. 329-380; see also Michèle Barrett, *The politics of truth: from Marx to Foucault* (Cambridge: Polity Press, 1991) pp. 96-110. For the relations between Lefebvre and Althusser, see Michael Kelly, *Modern French Marxism* (Oxford: Basil Blackwell, 1982) and Gregory, 'Modernity' (1994), pp. 355-7.

[2] For a summary, see Anthony Elliott, 'Psychoanalysis, ideology and modern societies: post-Lacanian social theory', in his *Social theory and psychoanalysis in transition: self and society from Freud to Kristeva* (Oxford, England and Cambridge, Mass: Blackwell Publishers, 1992) pp. 162-200, especially pp. 164-177.

[3] Lefebvre had dismissed Sartre's *L' Etre et le Néant,* published in 1943, in uncompromisingly hostile terms, but the publication of the first part of his *Critique de la raison dialectique* in 1960 marked both Sartre's philosophical acceptance of Marxism and Lefebvre's (still critical) acceptance of Sartre.

[4] It was in fact Foucault, another of Lefebvre's *bêtes noires,* who invited Lacan to conduct his seminar at Vincennes, after the École Normale Supérieure refused to allow him to continue in its own precincts. For a general (though hardly disinterested) discussion of the connections between French philosophy and the events of May 68, see Luc Ferry and Alain Renaut, *French philosophy of the sixties: an essay on anti-humanism* (Amherst: University of Massachusetts Press, 1990); this was originally published in French in 1985. Lefebvre provides his own account of those events in his *The explosion: Marxism and the French Revolution* (New York: Monthly Review Press, 1969); this was originally published in French in 1968.

sharpening those points which provoke Lefebvre into such vocal disagreement, however, it should be possible to develop a clearer sense of those other, counter-claims that he seeks to advance. It follows that my intentions in this essay are largely expository : I hope that by locating one of the unremarked bases from which Lefebvre constructs his history of space, a more vigorously critical appreciation of his work might be set in train.

Lacan's speculations

Lacan posits three «orders» - the Real, the Imaginary and the Symbolic - which together form a complex topological space in which, as Malcolm Bowie puts it, «the characteristic disorderly motions of the human mind can be plotted.» The spatial metaphoric becomes progressively more important in Lacan's work, which is what both attracts and annoys Lefebvre, and Lacan eventually configures this topological space as a Borromean chain (or «knot»), a complex figure formed from two separate links joined by a third in such a way that the chain will fall apart if any one of the links is severed.[1]

The Real

The Real is one of Lacan's most elusive concepts, and it also marks the site of one of Lefebvre's most fundamental disagreements with him. As a first approximation, one might say that the Real is «an anatomical, 'natural' order» into which a child is born and through which the child experiences its being as a «body-in-parts» : as a set of unco-ordinated, fragmented «raw materials» that Lacan calls, in an artfully gendered play on words, *un homelette*. This primal phase is soon organised through the other two orders, the Imaginary and the Symbolic, which correspond to subsequent developmental phases. But there is an interpretative difficulty here. Elizabeth Grosz treats the Lacanian Real as «a pure plenitude or fullness» exemplified by «the lack of a lack» - a reading which turns out to be much closer to Lefebvre's inclinations - whereas Bowie suggests that plenitude is approached by «reading off one by one the interferences between the Symbolic, Imaginary and Real by which «being human' is defined» : a claim which clearly accords with Lacan's topological representation of the three orders. Seen in this latter way, the Real is a permanent, «unrecoverable

---

[1] Malcolm Bowie, *Lacan* (London: Fontana, 1991) pp. 98-9; for examples of Lacan's topologies, see Jeanne Granon-Lafont, *La topologie ordinaire de Jacques Lacan* (Paris: Point hors Ligne, 1985) and Alexandre Leupin, 'Voids and knots in knowledge and truth', in Alexandre Leupin (ed.), *Lacan and the human sciences* (Lincoln: University of Nebraska Press, 1991) pp. 1-23.

presence» as it were, caught in a force field of tension and resisting its always unsuccessful re-presentation through the Imaginary and the Symbolic.[1]

It follows that the Real is not to be confounded with «reality» which is only lived and known through the Imaginary and the Symbolic. Although these other two orders were not developed in this way, they can I think be related to Lacan's invocation of Hegel's *Phenomenology* as a warrant for his own claim that the self can only grasp itself «through its reflection in, and recognition by, the other person.»[2] But Lacan is not prefiguring any Hegelian synthesis, and the ineluctable imperfections of that «grasping», and most particularly of attempts to conjure the Real into images and words, need to be stressed, because they help to explain why Lacan's attempts to delineate the Real prove to be so elusive. As Bowie reminds Lacan's frustrated readers, «allowing the structure of the Real to emerge against the background of a primitive, undifferentiated All is not the same thing as being able to name it, process it symbolically and put it to work for one's own ends.»[3]

The Imaginary

Lacan introduces the Imaginary through an allegory that turns on a developmental phase wholly absent from classical Freudianism : the so-called «mirror stage». Lacan claims that somewhere between the ages of six and eighteen months the young child typically develops a sense of wholeness, of bodily integrity and subjective unity, by looking at its own reflection in a mirror. He contrasts the child's «jubilant assumption of his *[sic]* specular image» to the response of a chimpanzee of the same age, who rapidly tires of playing with the mirror and moves off in search of other distractions. But the point of the comparison is not to celebrate human development. On the contrary :

> [S]omething derisory is going on in front of the mirror. Where the chimpanzee is able to recognise that the mirror-image is an epistemological void, the child has a perverse will to remain deluded. The child's attention is seized *(capté)* by the firm spatial relationships between its real body and its specular body and between body and setting within the specular image; he or she is captivated *(captivé)*. But the term that Lacan prefers to either of these, and which harnesses and outstrips their combined expressive power, is the moral and legal *captation :* the complex geometry of body, setting and mirror works upon the individual

---

[1] Elizabeth Grosz, *Jacques Lacan: a feminist introduction* (London: Routledge, 1990) pp. 33-4; Bowie, *Lacan* (1991), p. 99. I am indebted to Steve Pile for clarifying this interpretative puzzle for me.

[2] Anthony Elliott, 'The language of desire: Lacan and the specular structure of the self', in his *Social theory and psychoanalysis* (1992), pp. 123-161; the quotation is from p. 125. I should add that the two concepts were developed at different times: Lacan first presented his ideas about the Imaginary in 1936, though a printed (and revised) version did not appear until 1949, and the Symbolic was originally formalised during the 1950s and 60s.

[3] Bowie, *Lacan* (1991)., p. 95.

as a ruse, a deception, an inveiglement. The mirror, seemingly so consoling and advantageous to the infant, is a trap and a decoy *(leurre)*.[1]

This pre-verbal register constitutes the Imaginary, and I want to emphasise three features of Lacan's characterisation of it.

First, its inherent, constitutive *spatiality*. On this occasion at least, Lacan is offering more than a metaphor. As Jameson recognises,

> A description of the Imaginary will therefore on the one hand require us to come to terms with a uniquely determinate configuration of space - one not yet organized around the individuation of my personal body, or differentiated according to the perspectives of my own central point of view - yet which nonetheless swarms with bodies and forms intuited in a different way, whose fundamental property is, it would seem, to be visible without their visibility being the result of the act of any particular observer, to be, as it were, already-seen, to carry their specularity upon themselves like a colour they wear or the texture of their surface. In this ... these bodies of the Imaginary exemplify the very logic of mirror images; yet the existence of the normal object world of adult everyday life presupposes this prior, imaginary experience of space.[2]

Secondly, the *visuality* of the Imaginary. Lacan was extremely interested in Caillois's work on psychasthenia, in which the relation between the self and its surrounding space is disturbed through a visual fusion, a mimicry so complete that the self is assimilated to space. «To these dispossessed souls,» Caillois wrote, «space pursues them, encircles them, digests them in a gigantic phagocytosis.» But where Caillois described a crisis of the boundaried self, Lacan was interested in its opposite : in the formation of that self through specular identification.[3] For Lacan, Grosz argues, vision most readily confirms the separation of subject from object, and is also «the most amenable of the senses to spatialization.» Within the Imaginary, «space is hierarchically organized and structured in terms of a centralized, singularized point-of-view by being brought under the dominance of the visual.»[4]

Thirdly, its *duplicity*. Lacan plainly distrusts this spatialized, specularized self, which is founded on «a mirage of coherence and solidity through which the

---

[1] Bowie, *Lacan* (1991), p. 23. If celebration is in order, it would be difficult to think of a worthier subject than a chimpanzee that can recognise an 'epistemological void'.

[2] Jameson, 'Imaginary and Symbolic' (1977), pp. 354-5.

[3] Roger Caillois's original essay was published in *Minotaure* in 1935; it has been translated and reprinted as 'Mimicry and legendary psychasthenia', *October* 31 (1984) pp. 17-32; the quotation is from p. 30. My discussion also draws on Jay, 'The specular subject of ideology' (1994), pp. 342-3. I might add that psychasthenia provides another way of figuring Jameson's disorientation in postmodern hyperspace: see Celeste Olalquiaga, *Megalopolis: contemporary cultural sensibilities* (Minneapolis: University of Minnesota Press, 1922) pp. 1-5, 17-18.

[4] Grosz, *Jacques Lacan* (1990), p. 38. These considerations prompt Grosz to implicate Lacan in an ocularcentrism, but her discussion ignores the complex ways in which vision is gendered - even as she indicts Lacan for his phallocentrism - and forecloses Lacan's *critique* of ocularcentrism: see Jay, 'The specular subject of ideology' (1984), pp. 353-70.

subject is seduced into misrecognition of its own truth.» The key word is «misrecognition» - what Lacan called *méconnaissance* - which has been accentuated by virtually every commentator : thus «the capture of the 'I' by the reflection in the mirror is inseparable from a misrecognition of the gap between the fragmented subject and its unified image of itself.»[1] Taken together, therefore, one might say that, for Lacan, «the map of the body, setting and mirror both captivates and consoles the child, but it is an illusion, a trap, a decoy; [imaginary] geography is the medium of deception, it «offers 'ground truth' but cannot be trusted.»[2]

The Symbolic

What frees the subject from this hall of mirrors - or at any rate prevents its glissade into psychosis - is its passage into the Symbolic. This marks the child's entry into language, and Lacan's preoccupation with language is one of the signal elements of his own work. I mean by this something more than the analytical priority he accords to language, because Lacan's playfulness within language - in both his speech and his writing - is an integral part of his project. If «Lacan's writing seeks to tease and seduce,» as Bowie says, then its multiple «feints, subterfuges, evasions and mimicries» are (im)precisely what Lacan is talking *about,* (in)exactly what he has to *show :* if «the unconscious is structured like a language», then its irruptions cannot be other than plural, allusive, heterogeneous.[3]

With the installation of the Symbolic as the signifying register, Bowie argues, Lacan effectively turns the subject into a series of events within language : «the signifier becomes a versatile topological space, a device for plotting and replotting the itineraries of Lacan's empty subject.»[4] This imaginative analytical cartography has two roots (routes?) that reappear, in displaced form, in Lefebvre's spatial architectonics. One is inside the signifying chain and the other is inside the system of intersubjectivity, though these are articulated through one another, and I need to consider each of them in turn.

In the first place, Lacan turns to structural linguistics to identify two modes of connection within the signifying chain : *metaphor,* which is supposed to mark

---

[1] Peter Dews, 'Jacques Lacan: a philosophical rethinking of Freud', in his *Logics of disintegration: post-structuralist thought and the claims of critical theory* (London: Verso, 1987) pp. 45-86; the quotation is from p. 55; Elliott, 'Language of desire' (1992), p. 128.

[2] Steve Pile, 'Human agency and human geography revisited: a critique of "new models" of the self', *Transactions of the Institute of British Geographers* 18 (1993) pp. 122-139; the quotation is from p. 135.

[3] Bowie, *Lacan* (1991), p. 200. Bowie suggests that Lacan 'eroticizes' the language of theory, and interestingly Eagleton makes a similar point about Jameson:

> 'Discourse must be reinvested with desire, but not to the point where it confiscates the historical realizations of that desire. Jameson's style is a practice which displays such contradictions even as it strives to mediate them.'

See Terry Eagleton, 'Fredric Jameson and the politics of style', in his *Against the grain: essays 1975-1985* (London: Verso, 1986) p. 69.

[4] Bowie, *Lacan* (1991), p. 76.

the relation of discourse to the subject, and *metonymy,* which is supposed to mark the relation of discourse to the object. The importance of this manoeuvre is that it treats the unconscious as «the conjectural sub-text that is required in order to make the text of dreams and conversations intelligible.» It removes the occult quality of the unconscious - makes it accessible to analysis - by identifying these two rhetorical tropes with what Freud took to be the fundamental mechanisms of the dream-work : thus the signifying domain of the dream-work is decoded by identifying metaphor with «condensation» and metonymy with «displacement».[1] More than this, however, Lacan also implies an intricate gendering of the two processes. Metaphor's substitution of one word for another crosses the Saussurean bar between signifier and signified and can be read as an index of (phallic) «verticality», whereas metonymy's substitution of part for whole remains on one side of the bar and suggests a contiguity of femininity and «horizontality» : thus the object appears «beyond» or «on the other side of 'discourse'» as an absence, a lack, which is invested with *desire.*[2]

In the second place, and closely connected to these considerations, Lacan turns to structural anthropology, in particular to Lévi-Strauss's seminal discussion of the incest taboo, in order to conceptualise the Oedipus complex as a linguistic transaction. Most importantly, he identifies those agencies that place prohibitions on the child's desire for its mother with the «symbolic father» whose name supposedly initiates the liquid mobility of the signifying chain : with what Lacan calls the Name-of-the-Father, an ascription which in spoken French cleverly blurs the *«Non» du Père* (prohibition) with the *«Nom» du Père* (symbolization).[3] These prohibitions are inscribed within the Symbolic, and by this means Lacan equates entry into language with castration and identifies this loss or lack with the *phallus,* thus :

> [T]hrough his relationship to the signifier, the subject is deprived of something of himself, of his life, which has assumed the value of that which binds him to the signifier. The phallus is our term for the signifier of his alienation in signification. When the subject is deprived of this signifier, a particular object becomes for him an object of desire....[4]

I should say at once that Lacan's «phallus» cannot be immediately and directly identified with the penis - it connotes a signifier before a physical organ[5] - and

[1] Bowie, *Lacan* (1991), pp. 70-71.

[2] My understanding of these relations is indebted to Jane Gallop, 'Metaphor and metonymy' in her *Reading Lacan* (Ithaca: Cornell University Press, 1985) pp. 114-132. She is wonderfully successful in conjuring up co-existent and contradictory readings of metaphor and metonymy in Lacan, and in showing that the oppositions between the two terms and their connotations are far from straightforward or stable.

[3] Jay, 'The specular subject of ideology' (1984), p. 352. See also Grosz, *Jacques Lacan* (1990), pp. 101-105.

[4] These remarks are taken from Lacan's seminars on *Hamlet,* and are cited in Kaja Silverman, *The subject of semiotics* (New York: Oxford University Press, 1983) p. 183. As Lacan observes elsewhere, therefore, 'Nothing exists except on an assumed foundation of absence.'

[5] Though the two are by no means unconnected: see Kaja Silverman, 'The Lacanian phallus', in *Differences* 4 (1992) pp. 84-115.

that his primary point is about linguistic or symbolic castration. As Jane Gallop puts it, Lacan's Symbolic means that «we can only signify ourselves in a symbolic system that we do not command.»[1] And yet this is not a counsel of despair; the burden of Lacan's work - of his teachings and writings - is that it ought to be possible to comprehend our lack of comprehension, to learn from our loss, not to recover some (imaginary) plenitude but to come to terms with the tantalisation of language. Gallop's reading strategy turns on this very possibility : hence what one commentator criticises as her «insufficient command» of Lacan becomes a central part of her project as she struggles to disclose the ambiguities and evasions that enter into her «pathology of interpretation».[2] Delimiting language in this way, signalling its limits, is by not confined to the analyst or critic, however, and Lacan accentuates the reflexivity of all intersubjective action, thus :

> Although it may at first appear that the other subject is perpetually hidden «behind» the wall of language, it becomes apparent that all subjects are on the same side of this wall, although they are able to communicate only indirectly by means of the echo of their speech upon it.[3]

In general, then, one might conclude that Lacan's project issues in what Gallop calls «an implicit ethical imperative» to disrupt the Imaginary in order to reach the Symbolic. His forceful reconstruction of psychoanalysis works through a critique of *ocularcentrism* (of the Imaginary - a shattering of the mirror) which is at the same time far from being an assent to *logocentrism*. It is perfectly true that, on occasion Lacan seems to privilege the Symbolic over the Imaginary, to exult in the play of language and to break «the alibi of visual plenitude»[4]. But it is also true that he is immensely suspicious of all such binary oppositions, which he takes to indicate the continued presence and enduring power of the Imaginary : phrasing matters thus, privileging the one over the other, indicates the irruption of the Imaginary into the Symbolic.

---

[1] Gallop, *Reading Lacan* (1982), p. 20. This does not absolve Lacan of phallocentrism, however, and Gallop provides a constructively critical account in her 'Reading the Phallus', (1982), pp. 133-156; see also her *The Daughter's Seduction: feminism and psychoanalysis* (Ithaca: Cornell University Press, 1982). For further discussions, see Bowie, 'The meaning of the phallus' in *Lacan* (1991), pp. 122-157 and Jane Flax, *Thinking fragments: psychoanalysis, feminism and postmodernism in the contemporary West* (Berkeley: University of California Press, 1990) pp. 97-107.

[2] Gallop, *Reading Lacan* (1982), pp. 19-21, 131.

[3] Dews, 'Lacan' (1987), pp. 79-80.

[4] The term is from Jay, 'The specular subject' (1984), p. 359; Jay goes on to provide an important discussion of the difference (the split) that Lacan proposes between the eye and the gaze. For a particularly imaginative appropriation of Lacan, which speaks directly to Lefebvre's concerns, see Kaja Silverman, 'Fassbinder and Lacan: a reconsideration of the gaze, look and image', in her *Male subjectivity at the margins* (New York: Routledge, 1992) pp. 125-156.

Lefebvre's architectonics

In *La production de l'espace* Lefebvre refers directly to Lacan on only four occasions, always in footnotes and never in the body of the text. But his debt to Lacan is much greater than these few citations suggest, and in the introduction he sketches a spatial architectonics derived more or less immediately from Lacan :

> [O]ne might go so far as to explain social space in terms of a dual prohibition : the prohibition which separates the (male) child from his mother because incest is forbidden, and the prohibition which separates the child from its body because language in constituting consciousness breaks down the unmediated unity of the body - because ... the (male) child suffers symbolic castration and his own phallus is objectified for him as part of outside reality.[1]

This is not how Lefebvre constructs his own architectonics, let me say, but what is important about such a psychoanalytics of space, so he argues, is that it prepares the ground for an analysis of the spatial inscription of «phallic verticality» and horizontal partition. In his history of space he pays particular attention to the phallocentrism of abstract space and to the use of «walls, enclosures and façades ... to define both a *scene* (where something takes place) and an *obscene* area to which everything that cannot or may not happen on the scene is relegated.»[2] If Lefebvre's critique of Lacan is discriminating, however, this does not diminish its force. He is most critical of what he takes to be the imperialism of psychoanalysis, and insists that to explain everything in its terms «can only lead to an intolerable reductionism and dogmatism». But Lacan's work has a greater influence on Lefebvre's project than these comments imply, because he advances many of his own claims about the production of space through a critical dialogue with psychoanalysis. In many ways, I think that Lefebvre's response to the spatial architectonics sketched in the passage above provides the parameters for his own discussion. He objects that such a schema assumes «the logical, epistemological and anthropological priority of language over space» and that it puts prohibitions not productive activity at the heart of social space.[3] As I now want to show, his own project seeks to reverse these priorities; but in reversing them, his work carries forward, in displaced and distorted form, the same conceptual grid.

---

[1] Lefebvre, *Production* (1991), pp. 35-6.

[2] Lefebvre, *Production* (1991), p. 36; Lefebvre subsequently talks about a 'psychoanalysis of space' in exactly these terms (p. 99).

[3] Lefebvre, *Production* (1991), p. 36. The emphasis on *production* is vital, not because Lefebvre denies the reality of prohibition - he does not - but because it distances him from any preoccupation with 'space in itself' that would inculpate him in a spatial fetishism: p. 90. It also enables him to treat social space as 'not only the space of "no", [but] also the space of the body, and hence the space of "yes", of the affirmation of life': p. 201.

«An intelligence of the body»

Lefebvre's style is often as allusive as Lacan's, and he shares a similar exultation *in* language that is at the same time a suspicion *of* language. «Man does not live by words alone,» he declares : «In the beginning was the Topos», «long before ... the advent of the Logos.» What Lefebvre seeks to invoke by this cryptic phrase is a tensely organic spatiality, «an intelligence of the body», rooted in the taking place of practical activity and bound up with what, in a glancing blow at Lévi-Strauss's structural anthropology, he terms «the elementary forms of the appropriation of nature».[1] Lefebvre talks of a time, long before the inauguration of either «historical» or modern, «abstract» spatialities, when the body's relationship to space had «an immediacy which would subsequently degenerate and be lost.» He grounds this «absolute space» in «a biologico-spatial reality», but it is plainly not a projection of the Lacanian Real onto the plane of society : there is an implication not only of intimacy but of plenitude, inseparable from the dense figuration of absolute space in practices of measurement and representation that «held up to all members of a society an image and a reflection of their own bodies.»[2] Neither is this a projection of the Lacanian Symbolic onto the plane of society; these «images» and «reflections» were not separate from the corporeality of the body or the physicality of the natural world but were fully continuous with them. They formed analogical and cosmological spaces within an absolute space whose meanings were not assigned to some separate symbolic register : they belonged to a world in which, as Lefebvre displaces Lacan, «the imaginary is transformed into the real.»[3]

The distorted echoes of Lacan in that phrase are unmistakable, but I propose to sharpen Lefebvre's argument (and to clarify those distortions) by means of a different comparison. Michel Foucault had already suggested that the human body «is always the possible half of a universal atlas.» He illustrated the claim with a series of vivid vignettes, many of which were unknown to or even unrecognised by some of the astonished historians who read *Les mots et les choses.* For example :

> Upright between the surfaces of the universe, [man] stands in relation to the firmament (his face is to his body what the face of heaven is to the ether; his pulse beats in his veins as the stars circle the sky according to

---

[1] Lefebvre, *Production* (1991), pp. 117, 174. Cf. Claude Lévi-Strauss, *Les structures élémentaires de la parenté* (Paris: Presses Universitaires de France, 1949) translated into English as *The elementary structures of kinship* (Boston, Mass: Beacon Press, 1969). Lévi-Strauss's relationship to historical materialism was always contentious which is presumably why Lefebvre invokes classical Marxism through 'the appropriation of nature.' Lefebvre subsequently confronts Lévi-Strauss directly: his structural anthropology is another strategy of abstraction, in which space becomes merely 'a means of classification, a nomenclature for things, a taxonomy' quite independent of their content. Lefebvre is also and in consequence astonished at Lévi-Strauss's determination to discuss kinship without discussing sexuality, eroticism or desire: Lefebvre, *Production*, p. 296.

[2] Lefebvre, *Production* (1991), pp. 110-1.

[3] Lefebvre, *Production* (1991), p. 251. For a fuller discussion of the distinctions between the analogical and cosmological spaces of absolute space and the emergence of the symbolic spaces of historical space, see Gregory, 'Modernity' (1994) pp. 382-92.

their own fixed paths; the seven orifices in his head are to his face what the seven planets are to the sky); but he is also the fulcrum upon which all these relations turn, so that we find them again, their similarity unimpaired, in the analogy of the human animal to the earth it inhabits : his flesh is a glebe, his bones are rocks, his veins great rivers, his bladder is the sea, and his seven principal organs are the metals hidden in the shafts of mines.[1]

«Man», Foucault wrote, «transmits these resemblances back into the world from which he receives them.» And for Foucault - though not Lefebvre - «the space inhabited by immediate resemblances becomes like *a vast open book»* : «The great untroubled mirror in whose depths things gazed at themselves and reflected their images back to one another is, in reality, filled with *the murmur of words.*»[2] To be sure, Foucault was describing the episteme of sixteenth-century Europe (or so he said), whereas Lefebvre conjures up an altogether different culture in time and space. But these distinctions are more than historico-geographical markers. If the «great mirror» of Lefebvre's absolute space also reveals «the prose of the world», if its language has also «been set down in the world and forms part of it» - I take these phrases from Foucault - it is emphatically not textualised in the same way and does not privilege the written word.[3] Lefebvre acknowledges that absolute space was marked by various means, and concedes that in this sense one might say that «practical activity writes upon nature». But he prefers to speak of the production of *textures* not texts. Analogies between space and language space «can only be carried so far», he warns, because space is *«produced* before being *read»,* marked by demarcations and orientations not in order to be read - deciphered - but «rather in order to be *lived* by people with bodies.»[4]

　　Lefebvre's history of space reconstructs the degradation of this organic spatiality, and it should already be apparent that there is something nostalgic

---

[1] Michel Foucault, *Les mots et les choses* (Paris: Gallimard, 1966) translated into English as *The order of things: an archaeology of the human sciences* (London: Tavistock, 1970) p. 22. The original French publication coincided with that of Lacan's *Écrits;* Foucault had heard Lacan lecture at the École Normale Supérieure, and psychoanalysis was one of the 'counter-sciences' he discussed in *Les mots et les choses.* Didier Eribon claims that 'Foucault's entire archaeological enterprise was really based on Lacan', but his other biographers - David Macey and James Miller - contend that Foucault had little sympathy with Lacan's project. Certainly, by the time of *La volonté de savoir,* the first volume of his projected *Histoire de la sexualité,* even Eribon agrees that Foucault was 'setting out on a genealogical quest against Lacan': see Didier Eribon, *Michel Foucault* (Cambridge, Mass: Harvard University Press, 1991) p. 272; this was originally published in French in 1989. Lacan's close collaborator (and son-in-law) has provided a brief post-mortem discussion of the importance of psychoanalysis in Foucault's writings: see Jacques-Alain Miller, 'Michel Foucault and psychoanalysis', in Timothy Armstrong (trans.), *Michel Foucault: Philosopher* (London: Routledge, 1992) pp. 58-63.

[2] Foucault, Order of things (1992) p. 27 (my emphases).

[3] Foucault (1992), pp. 34-5, 38-9. Lefebvre talks of 'the prose of the world' too, but he has in mind *buildings* which constitute 'the homogeneous matrix of capitalist space' and which he reads within the symbolic dimension of spatiality. See Lefebvre, *Production* (1991), p. 227.

[4] Lefebvre, *Production* (1991), pp. 118, 132, 143. Lefebvre's distance from Foucault (here and elsewhere) is thus considerable. What interests Lefebvre is not space *as* language therefore, or landscape *as* text, but rather 'as-yet concealed relations between space *and* language' (p. 17; my emphasis).

about his project. Its characteristic conjunction of humanism and romanticism faces in quite the opposite direction to Lacan. Where Lefebvre offers a threnody for a lost plenitude, and in places seems to hold out the hope that «authenticity» and subjectivity can be recovered through a kind of corporeal mimesis, Lacan's Real always already resists representation and recovery : as Merquior sees it, from a position that is not dissimilar to Lefebvre's, Lacan is «the archdenier of fulfillment, for whom all desire was bound up with the tragedy of lack and dissatisfaction.»[1] Lefebvre yearns for a return to his «real», to the fulfillment of what Marx once described as «man in the whole wealth of his being», whereas Lacan bars any such return : that «wealth» (if I can extend the metaphor) is invested in all three orders, Real, Imaginary and Symbolic, and human being necessarily capitalises on all three.

The tension between the two thinkers is compounded further because Lefebvre treats the degradation of this organic spatiality as at once «developmental» and «historical», and his discussion slips in and out of these different temporalities. This makes exposition difficult, partly because these slippages mark sites of ambiguity (and sometimes, I suspect, evasion), but partly because Lacan's treatment is conducted within the developmental register alone. For Lefebvre, however, the politics of space are inseparable from both the history of the body and the history of space and, as I have said, he pays particular attention to the formation of abstract space through an historical process of decorporealization (which thus becomes the dual of «abstraction»). Although my continued presentation of Lefebvre will parallel my presentation of Lacan, for these reasons I hope it will be understood if on occasion it becomes necessary to follow tangents which, in some places, will ineluctably turn from clear lines of flight into tangled labyrinths.

«The fruits of dread»[2]

I resume with Lefebvre's own discussion of the mirror, which is clearly provoked by Lacan (and, indirectly, by Althusser's reading of Lacan). Lefebvre complains that psychoanalysis plays too freely with the mirror effect and abstracts it «from its properly spatial context, as part of a space internalized in the form of mental 'topologies'».[3] This is an objection which, in its more general form, provides the original spur to Lefebvre's project : the inability of other thinkers to eschew «the basic sophistry whereby the philosophico-epistemological notion of space is fetishized and the mental realm comes to

---

[1] J. G. Merquior, *From Prague to Paris: a critique of structuralist and post-structuralist thought* (London: Verso, 1986) p. 155. I yoke Merquior to Lefebvre through their antipathy to structuralism and post-structuralism; Merquior treats Lacan as the principal architect of the bridge between structuralism and post-structuralism (p. 149).

[2] The phrase is Tristan Tzara's and refers to mirrors: see Lefebvre, *Production* (1991),p. 184n.

[3] Lefebvre, *Production* (1991, p. 184n.

envelop the social and physical ones.»[1] In Lacan's case, Lefebvre remarks that «the mirror helps to counteract the tendency of language to break up the body into pieces, but it freezes the Ego into a rigid form rather than leading it towards transcendence in and through a space which is at once practical and symbolic (imaginary).»[2] The elision between «imaginary» and «symbolic» is obviously problematic if these terms are understood in a strictly Lacanian sense, but that is precisely what Lefebvre intends : he wants to establish the double inscription of the two in social space and the materiality of that conjunction. «Into that space which is produced first by natural and later by social life the mirror introduces a truly dual spatiality : a space which is imaginary with respect to origin and separation, but also concrete and practical with respect to co-existence and differentiation.»[3]

Let me explain this double inscription. Lefebvre begins by using the mirror and the mirror-effect to suggest the intimate involvement of the production of social space in the constitution of the self :

> On the one hand, one ... relates oneself to space, situates oneself in space. One confronts both an immediacy and an objectivity of one's own. One places oneself at the centre, designates oneself, and uses oneself as a measure.... On the other hand, space serves as an intermediary or mediating role : beyond each plane surface, beyond each opaque form, «one' seeks to apprehend something else. This tends to turn social space into a transparent medium occupied solely by light, by «presences» and influences.[4]

The production of social space as duality has profound consequences, for Lefebvre suggests that this shift from «the space of the body to the body-in-space» facilitates the «spiriting-away» or *scotomization* of the body. The term originated in opthalmology in the nineteenth century, where it was used to refer to a lacuna in the field of vision, and had been introduced into French psychoanalytic theory in the 1920s; it was later appropriated by Lacan to diagnose a psychosis produced by a particular entanglement of the visual and the linguistic. Lefebvre uses the term in an historical rather than a developmental sense but, as I will show, retains that imbrication of the visual and the linguistic.[5]

Lefebvre argues that space, which was originally known, marked and produced through all the senses - taste, smell, touch, sound and sight - and which was, in all these ways, in intimate conjunction with the «intelligence of the body», comes to be constituted as a purely visual field. He represents this

---

[1] Lefebvre, *Production* (1991), p. 5. Lefebvre's indictment includes Althusser, Barthes, Derrida, Foucault, Kristeva and Lacan.

[2] Lefebvre, *Production* (1991), p. 185n.

[3] Lefebvre, *Production,*(1991), p. 186.

[4] Lefebvre, *Production* (1991), pp. 182-3.

[5] Lefebvre, *Production* (1991), p. 201; Jay, 'The specular subject of ideology' (1984), pp. 353-7. I am summarizing and simplifying a complicated argument about scotomization here; for a fuller discussion, see Carolyn Dean, *The self and its pleasures: Bataille, Lacan and the history of the decentred subject* (Ithaca: Cornell University Press, 1992).

process as a generalisation of the mirror-effect, in which social space itself becomes a collective mirror. Unlike Lacan, however, the importance of the mirror for Lefebvre is not that its reflection «constitutes my unity *qua* subject» but rather that «it transforms what I am into the *sign* of what I am.» In much the same way, therefore, Lefebvre suggests that within the «symbolic imaginary» of social space «the sign-bearing 'I' no longer deals with anything but other bearers of signs.» In effect, «space offers itself like a mirror to the thinking 'subject', but after the manner of Lewis Carroll, the 'subject' passes through the looking-glass and becomes a lived abstraction.»[1] This collective - and historical - passage marks the transformation from absolute into abstract space.

By the time this process is complete, space has no social existence independently of an intense, aggressive and repressive visualization. It is thus - not symbolically but in fact - a purely visual space. The rise of the visual realm entails a series of substitutions and displacements by means of which it overwhelms the whole body and usurps its role.[2]

Lefebvre maps what he calls this «logic of visualization» along two dimensions. The first is *metonymy,* which he treats as an axis of *spectacularization* by means of which «the eye, the gaze, the thing seen, no longer mere details or parts, are now transformed into the totality.» As vision asserts its primacy over the other senses, therefore,

All impressions derived from taste, smell, touch and even hearing first lose clarity, then fade away altogether, leaving the field to line, colour and light. In this way, a part of the object and what it offers comes to be taken for the whole.[3]

The second is *metaphor*, which he treats as an axis of *textualization* by means of which words substitute for images and the conduct of social life becomes synonymous with «the mere reading of texts» :

Living bodies, the bodies of «users» ... are caught up not only in the toils of parcellized space, but also in the web of what philosophers call «analagons» : images, signs and symbols. These bodies are transported out of themselves, transferred and emptied out, as it were, via the eyes.[4]

---

[1] Lefebvre, *Production* (1991), pp. 185 (my emphasis), 311, 313-4.

[2] Lefebvre, *Production* (1991), p. 286.

[3] Lefebvre, *Production* (1991), p. 286. In speaking of 'spectacularization' Lefebvre is acknowledging his debt to his (once close) collaboration with the Situationist International and, in particular, to Guy Debord, *La société du spectacle* (Paris: Éditions Buchet-Chastel, 1967); this has been translated into English as *The society of the spectacle* (Detroit: Black and Red, 1983). For an imaginative commentary on the periodization of the spectacle, which intersects with Lefebvre's thesis in suggestive ways, see Jonathan Crary, 'Spectacle, attention, counter-memory', *October* 50 (1989) pp. 99-107.

[4] Lefebvre, *Production* (1991) p. 98. For a salutary commentary on explicitly spatial metaphors, which takes its cue directly from Lefebvre, see Neil Smith and Cindi Katz, 'Grounding metaphor: towards a spatialized politics', in Michael Keith and Steve Pile, *Place and the politics of identity* (London: Routledge, 1993) pp. 67-83. Their central point is Lefebvre's: 'Spatial metaphors are problematic in so far as they presume space is not.'

The parallels and tangents, resonances and dissonances between Lefebvre and Lacan could scarcely be clearer, but Lefebvre prefers to attribute these ideas to Nietzsche. It was Nietzsche who disclosed the historical advance of the visual over the other senses, he notes, and Nietzsche who identified «the visual aspect predominant in the metaphors and metonyms that constitute abstract thought.»[1] The acknowledgement is significant, not only because it gestures towards Nietzsche's general importance to Lefebvre's project but also because it accentuates the historical specificity of his own account of metaphor and metonymy. He is scathing about those appropriations of psychoanalysis that provide «homogenizing» explanations which cannot comprehend diversity. In his reflections on «monuments» and «buildings», for example, he treats these linguistic operators in ostensibly Lacanian terms and identifies the condensations and displacements which they effect; but he also insists that the Saussurean bar separating «signifier from signified and desire from its object» is placed differently in different social spaces and that these concepts will not explain very much until they «address the question of which particular power is in place.»[2] For this reason, it is necessary for me to emphasise that the foregoing discussion concerns the metaphorical and metonymical dimensions of *abstraction* - the materials they work with and the outcomes of their operations in other process-domains will be different - and that Lefebvre is using «abstraction» in a highly specific way.

Still, notice what has happened : Lefebvre has used the mirror to conjure up what, for Lacan, would be the signifying order. By inscribing the imaginary and symbolic within the same social space, Lefebvre is able to register a double suspicion. First, he warns against *ocularcentrism* :

> Wherever there is illusion, the optical and visual world plays an integral and integrative ... part in it. It fetishizes abstraction and imposes it as the norm. It detaches the pure form from its impure content - from lived time, everyday time, and from bodies with their opacity and solidity, their warmth, their life and their death.[3]

This is at once a developmental and an historical thesis, I take it, and it underwrites Lefebvre's claim that the production of space - most particularly of abstract space - is typically concealed by a (double) illusion of transparency and opacity. Edward Soja has since used and extended these tropes to identify two illusions in the conventional theorization of spatiality : a «hypermetropic illusion» that «sees right through the concrete spatiality of social life by projecting its production into an intuitive realm of purposeful idealism and immaterialized reflexive thought» and a «myopic illusion» that produces «short-sighted interpretations of spatiality which focus on immediate surface appearances without being able to see beyond them.» These are important and insightful remarks, but Lefebvre's original claim was not a disciplinary or even a

---

[1] Lefebvre, *Production* (1991) p. 139.

[2] Lefebvre, *Production* (1991) pp. 225-227, 248.

[3] Lefebvre, *Production* (1991) p. 97.

trans-disciplinary one. These illusions are by no means confined to the academy - to «theory» - but enter into the material constitution of social space : into what he calls «the world-as-fraud».[1]

Secondly, Lefebvre is wary of *logocentrism*. He distinguishes two orientations toward language. The one is derived from semiotics and structuralism, and involves the exorbitation of language : in its most extreme form it becomes a formalism in which «everything - music, painting, architecture - is language» - and which is therefore directed to the salvation of «the Logos». The other is less elegiac, more tragic : it dwells on the brooding menace of the sign, the «intimate connection between words and death», and discloses «the secret of the Logos as foundation of all power and authority.»[2] Lefebvre's own sympathies are much closer to the second :

> This is a lethal zone thickly strewn with dusty, mouldering words. What slips into it is what allows meaning to escape the embrace of lived experience, to detach itself from the fleshly body. Words and signs facilitate (indeed provoke, call forth and - at least in the West - command) metaphorization - the transport, as it were, of the physical body outside of itself. This operation, inextricably magical and rational, sets up a strange interplay between (verbal) disembodiment and (empirical) re-embodiment, between uprooting and reimplantation, between spatialization in an abstract sense and localization in a determinate expanse. This is the «mixed space» - still natural, yet already produced - of the first year of life, and, later, of poetry and art. The space, in a word, of representations.[3]

The elision between developmental and historical temporalities is particularly stark in this passage. Lefebvre is implicitly addressing Lacan, and he prefaces these remarks with the suggestion that the unconscious be treated not as «a source of language» but rather as an interstice between the self-seeking-to-constitute-itself and its body; what then «slips into the interstice» is «language, signs, abstraction - all necessary yet fateful indispensable yet dangerous.» But he is also prefiguring an historical thesis, in which the production of space is marked by a progressive and aggressive logocentrism. «The Logos makes inventories, classifies, arranges : it cultivates knowledge and presses it into the service of power.»[4]

As I have indicated, then, these processes of spectacularization and textualization are implicated in the historical production of an *abstract space*. The term is at once entirely accurate and yet thoroughly deceptive. «Abstraction passes for an 'absence',» Lefebvre cautions, «as distinct from the concrete

---

[1] Lefebvre, *Production* (1991) pp. 27-30, 389; Soja, *Postmodern geographies* (1989) pp. 122-126.

[2] Lefebvre, *Production* (1991) pp. 133-135; I have borrowed 'the exorbitation of language' from Anderson, *Tracks* (1983) pp. 40-41. It seems particularly appropriate here: although Lefebvre represents psychoanalysis as the (dishonest) broker between these two positions, it seems clear that - like Anderson - he identifies Lacan with the first position and pays little attention to what Lacan also has to say from the second position.

[3] Lefebvre, *Production* (1991), p. 203.

[4] Lefebvre, *Production* (1991), pp. 203, 391-2.

'presence' of objects, of things», but «*nothing could be more false.*» He says this precisely because abstract space is not a void. In his view, «signs have something lethal about them» not only because they involve «devastation and destruction» - in particular, scotomization - but also because *they put in place an apparatus of repression.* This «visual space of transparency and readability», produced under the sign(s) of capitalist modernity, is one in which «exchanges between knowledge and power, and between space and the discourse of power, multiply and are regularized.» To forestall misunderstanding, I should say at once that Lefebvre's critique is not limited to the construction of quasi-Foucauldian disciplinary spaces, as this phrasing might imply, to the normalization of a space in which nothing escapes surveillance and where the theoretical and everyday practices of «reading space» are implicated in discursive technologies of *assujetissement.* These are important matters, to be sure, and Lefebvre does not neglect them (though he does not treat them in the same way as Foucault).[1] But Lefebvre also insists that the violence of abstraction - flattening, plan(n)ing, emptying space - at once unleashes and lures *desire.* «It presents desire with a 'transparency' which encourages it to surge forth in an attempt to lay claim to an apparently clear field.»[2]

The invocation of desire is immensely important, even though Lefebvre does not develop its topographies in any detail, because it means that abstract space cannot be evacuated altogether. On the contrary : «It demands a truly full object», so Lefebvre argues, a signifier «which, rather than signifying a void, signifies a plenitude of destructive force - an illusion, therefore, of plenitude, and a space taken up by an 'object' bearing a heavy cargo of myth.»[3] This turns out to be the (Lacanian) phallus writ large and physically inscribed in abstract space :

> Metaphorically, it symbolizes force, male fertility, masculine violence. Here again the part is taken for the whole [metonymy]; phallic brutality does not remain abstract, for it is the brutality of political power, of the means of constraint : police, army, bureaucracy. Phallic erectility bestows a special status on the perpendicular, proclaiming phallocracy as the orientation of space, as the goal of the process - at once metaphoric and metonymic - which instigates this facet of spatial practice.[4]

If this is a starker architecture than Lacan had in mind, if its materialism seems too crude in places, this may be because Lefebvre offers only illustrative sketches, signs indicating the sites at which a more subtle argument needs to be developed. But the nature of that project should now be clear : Lefebvre is bringing together three key elements from Lacan - the Eye, the Logos and the Phallus - and using them to gloss, to illuminate (in highly particular ways) the

---

[1] Lefebvre, *Production* (1991), pp. 282, 289; cf. Michel Foucault, *Discipline and punish: the birth of the prison* (London: Allen Lane, 1977); this was originally published in French in 1975.

[2] Lefebvre, *Production* (1991), p. 97.

[3] Lefebvre, *Production* (1991), p. 287.

[4] Lefebvre, *Production* (1991), p. 287.

historical production of abstract space. As he puts it with a characteristic flourish, «King Logos is guarded on the one hand by the Eye» - what he calls, in the same play of words that captivates Lacan, «the eye of the Father» - and on the other hand by the Phallus. Together these mark, represent and produce «the space of the written word and the rule of history.»[1]

Lefebvre works with these ideas to both disclose and contest a dominant phallocentrism. He does so through a double argument. In the first place, he constantly accentuates the gendering of Lacan's conceptual trinity and hence establishes the masculinism that is written into this abstract space.

> In abstract space, ... the demise of the body has a dual character, for it is at once symbolic and concrete : concrete, as a result of the aggression to which the body is subject; symbolic, on account of the fragmentation of the body's living unity. This is especially true of the female body, as transformed into exchange value, into a sign of the commodity and indeed into a commodity *per se*....[2]

Lefebvre repeatedly points to the conjunction between the deceptive «transparency» of abstract space and a masculinist violence, and there are poignant connections between his troubled attempt to come to terms with the symbolic imaginary of this space and Gillian Rose's critique of the masculinism of the geographical imagination in which, in theory and in practice, «the innocent transparency of the empty street» becomes, for her and for so many other women, «an aggressive plastic lens pushing on me.»[3] In the second place, Lefebvre tries to show that phallocentrism is so hideously destructive of what it is to be a human being that it has become one of the most powerful media of alienation. «Over abstract space reigns phallic solitude and the self-destruction of desire,» he writes, and its violent severance from the plenitude of the Real is equivalent to castration :

> Abstract space is doubly castrating : it isolates the phallus, projecting it into a realm outside the body, then fixes it in space (verticality) and brings it under the surveillance of the eye. The visual and the discursive are buttressed (or contextualized) in the world of signs.[4]

---

[1] Lefebvre, *Production* (1991), pp. 262, 408.

[2] Lefebvre, *Production* (1991), p. 310.

[3] Gillian Rose, *Feminism and geography: the limits of geographical knowledge* (Cambridge: Polity Press, 1993) p. 143. I do not mean to imply that their projects are coincident, however, and neither do I think these affinities protect Lefebvre's project from a feminist critique.

[4] Lefebvre, *Production* (1991), p. 310. Elsewhere Lefebvre refers to 'The Great Castration' (p. 410) which is presumably a play on Foucault's treatment of 'The Great Confinement.'

Histories, archaeologies

Although I have tried to show how Lefebvre's account of the production of space can be illuminated through a consideration of Lacan's psychoanalytic theory, this is plainly not the only way to read it. *La production de l'espace* is constructed from many different materials and there is no single route through all of its passages. One of the most telling differences from Lacan, as I have repeatedly remarked, is that Lefebvre offers a *history* that stands in close, critical proximity to *historical* materialism. In Lefebvre's view, psychoanalysis had a fatal flaw : «It had a non-temporal view of the causes and effects of a society born in historical time.»[1] The originality of his own work, within Western Marxism, lies in the connection it proposes between the history of space and the history of the body, however, and it is for precisely this reason that I think an approach through Lacan can be so revealing.[2] But it can also be concealing, and the historicity of Lefebvre's project has three implications that need to be underscored.

First, Lefebvre is adamant that the decomposition of the human body - what he calls the «body without organs» - and the decorporealization of social space cannot be «laid at the door of language alone». In his view, such a manoeuvre would wrongly exonerate the Judaeo-Christian tradition, «which misapprehends and despises the body, relegating it to the charnel-house if not to the Devil», and the capitalist mode of production and its advanced division of labour, which «has had as much influence as linguistic discourse on the breaking-down of the body into a mere collection of unconnected parts.»[3] It is for this reason that Lefebvre turns to historical materialism, of course, and en route he turns it into something very like David Harvey's historico-geographical materialism. Very like, but scarcely the same. Harvey's recent writings on postmodernity have clearly been inspired by Lefebvre's work, and he has had much to say about images, mirrors and representations; but he has yet to engage with psychoanalysis (except through Jameson's appeal to Lacan, and then only in passing).[4]

Secondly, there is undoubtedly a sense in which Lefebvre's long history of space treats the aggressive advance of a «logic of visualization» as a meta-

---

[1] Henri Lefebvre, *The survival of capitalism* (London: Allison and Busby, 1976) p. 31; this was originally published as *La survie du capitalisme* (Paris: Editions Anthropos, 1973). This may well be true of Lacan, and there is undoubtedly a universalizing temper to much of psychoanalytic theory, but such a claim ought not to become a misleading universal in its own right. Kristeva, for example, makes it very clear that the 'signifying economy' operates through the *biographical* subject and recasts him/her as an *historical* subject.

[2] Lefebvre, *Production* (1991), p. 196.

[3] Lefebvre, *Production* (1991), p. 204.

[4] David Harvey, *The condition of postmodernity: an enquiry into the origins of cultural change* (Oxford and Cambridge, Mass: Blackwell Publishers, 1989) pp. 53-4. Cf. Meaghan Morris, 'The man in the mirror: David Harvey's "Condition" of Postmodernity', *Theory, culture and society* 9 (1992) pp. 253-279. For a detailed discussion of other connections and contrasts between Lefebvre's project and Harvey's historico-geographical materialism, see Gregory, 'Modernity' (1994)

narrative : but I believe this is offered for strategic not stipulative purposes. He does not claim that the decorporealization of space is the only important thematic, nor even that it is the one which most inclusively gathers in the other possible narratives and *petits recits,* but simply that it is one which has been neglected and which has a vital, terrible salience for any critical politics of space. This also allows Lefebvre to speak to questions of gender, sexuality and phallocratic violence - and to their inscriptions in social space - more directly than many other writers of his time. Of course, the logic of visualization was not uncontested, and there were multiple attempts to interrupt or unravel its metanarrative. Lefebvre says little enough about them in concrete terms and prefers to remain at a metaphilosophical level. But he makes much of the «great struggle» between what Nietzsche called «the Logos and the Anti-Logos», between what he himself sees as «the explosive production of abstract space and the production of a space that is *other»,* and his own critique is intended as a challenge to the hegemony of that awful metanarrative.[1] Lefebvre's insistence on contestation is a political claim, deliberately so, and his project locates the bases for resistance - for the construction of authentically human spaces - in the past as well as the present. He is offering, or at least his writings can be made to offer, a «history of the present» in something like the sense in which Foucault employs the term. John Rajchman suggests that «a history of the present» has two connotations. In the first place,

> The «present» refers to those things that are constituted in our current proceedings in ways we don't realize are rooted in the past, and writing a «history» of it is to lay bare that constitution and its consequences.

Lefebvre's project can certainly be read in this way : the very idea of a history of space has long been ignored by many writers and critics. In the second place, however,

> Foucault does not show our situation to be a lawlike outcome of previous ones, or to have been necessitated by the latest historical conjuncture. On the contrary, he tries to make our situation seem less «necessitated» by history, and more peculiar, unique or arbitrary.[2]

In so far as Lefebvre's history of space is a historicism, then it is of necessity much less «singularizing» than this. But it also provides, on occasion, what Benjamin once called a constellation - I can think of no better term - in which it

---

[1] Gregory, 'Modernity' (1994), p. 391. Even at this general level, within the realms of reflection and speculation, it is possible to provide more detailed counter-histories. I have already drawn on Martin Jay's reconstruction of a persistent critique of ocularcentrism within twentieth-century French philosophy, for example, and I should also like to draw attention to Rosalind Krauss's wonderful reconstruction of art history and her evocation of a counter-modernism that inscribed the corporeality of vision within and against the power of the abstracted eye. See Jay, *Downcast eyes* (1993) and Rosalind Krauss, *The optical unconscious* (Cambridge, Mass: MIT Press, 1993).

[2] John Rajchman, *Michel Foucault: the freedom of philosophy* (New York: Columbia University Press, 1985) p. 58.

becomes possible for a (forgotten, even repressed) past to irrupt explosively into the present.

Thirdly, therefore, and following directly from this claim, Lefebvre situates his history of space somewhere between anthropology and political economy. His «anthropology» is the site of another awkward slippage between the developmental and the historical. Lefebvre suggests that the «history» of space is inaugurated as the plenitude of a «biologico-spatial reality» is lost, and he evidently believes that non-modern societies and their spaces are somehow closer to «the intelligence of the body» and its organic spatiality. But he also argues that those biomorphic resonances persist into our own troubled present and provide the distant murmurs of a fully human future :

> Nothing disappears completely... nor can what subsists be defined solely in terms of traces, memories or relics. In space, what came earlier continues to underpin what follows. The preconditions of social space have their own particular way of enduring and remaining actual within that space... The task of *architectonics* is thus to describe, analyse and explain this persistence, which is often invoked in the metaphorical shorthand of strata, periods, sedimentary layers...[1]

This is, in part, what Lefebvre's «rhythm analysis» was supposed to recover; it would seek to discover those rhythms, those circuits of plenitude as it were, «whose existence is signalled only through mediations, through indirect effects or manifestations.» He claimed that their presence is registered in the space of dreams, «where dispersed and broken rhythms are reconstituted», and for this reason he wondered whether rhythm analysis might not «eventually even displace psychoanalysis.» This turned out to be a vain hope, and it is in any case far from clear, even in Lefebvre's own exposition, what he had in mind.[2] But these muffled rhythms are also registered in the modern sphere of «everyday life», in the day-to-day practices and representations of modernity, which Lefebvre treats as both «a parody of lost plenitude and the last remaining vestige of that plenitude.»[3] This may be a more promising avenue of inquiry, though its elegiac romanticism remains awkward, but in any event it seems clear that his project is a thoroughly redemptive one. It should also be clear that his «rescue archaeology» is not Foucault's; neither is it Lacan's : it is, distinctively, Lefebvre's.

---

[1] Lefebvre, *Production* (1991), p. 229.

[2] Lefebvre, *Production* (1991), pp. 205, 209. See also Henri Lefebvre, *Éléments de rythmanalyse: introduction à la connaissance des rhythmes* (Paris: Éditions Syllepse, 1992).

[3] Michel Trebitsch, 'Preface' to Henri Lefebvre, *Critique of everyday life.* Vol. 1: *Introduction* (London: Verso, 1991) p. xxiv; this was first published as *Critique de la vie quotidienne* (Paris: Grasset, 1947). For a discussion of Lefebvre's account of the 'colonisation of everyday life' and the practical recovery of that 'lost plenitude', see Gregory, *Imaginations* (1994), pp. 401-406.

Acknowledgements: I have relied even more than usual on the comments and advice of friends, and it is a pleasure to acknowledge the help of Trevor Barnes, Alison Blunt, Noel Castree, Mike Crang, Chris Philo, Geraldine Pratt, Rose Marie San Juan, Matt Sparke, and Bruce Willems-Braun. I am particularly grateful to Steve Pile who commented in wonderfully constructive ways on a draft of the argument and generously shared his insights into Lacan's work with me.

References

Anderson, P., 1979, *Considerations on Western Marxism*, London: Verso

Anderson, P., 1983, *In the tracks of historical materialism*, London: Verso

Barrett, M., 1991, *The politics of truth: from Marx to Foucault,* Cambridge: Polity Press

Bowie, M., 1991, *Lacan*, London: Fontana

Bukatman, S., 1993, *Terminal identity: the virtual subject in post-modern science fiction,* Durham: Duke University Press

Butler, J., 1987, *Subjects of desire: Hegelian reflections in twentieth-century France*, New York: Columbia University Press

Caillois, R., 1984, 'Mimicry and legendary psychasthenia', *October* 31, 17-32.

Cohen, M., 1993, *Profane Illumination: Walter Benjamin and the Paris of surrealist revolution* Berkeley: University of California Press

Crary, J., 1989, 'Spectacle, attention, counter-memory', *October* 50, 99-107.

Dean, C., 1992, *The self and its pleasures: Bataille, Lacan and the history of the decentred subject*, Ithaca: Cornell University Press

Debord, G., 1983, *The society of the spectacle*, Detroit: Black and Red

Descombes, V., 1980, *Modern French Philosophy*, Cambridge: Cambridge University Press

Dews, P., 1987, *Logics of disintegration: post-structuralist thought and the claims of critical theory,* London: Verso

Dowling, W., 1984, *Jameson, Althusser, Marx*, Ithaca: Cornell University Press

Eagleton, T., 1986, *Against the grain: essays 1975-1985*, London, Verso

Elliott, A., 1992, *Social theory and psychoanalysis in transition: self and society from Freud to Kristeva*, Oxford, England and Cambridge, Mass: Blackwell Publishers

Eribon, D., 1991, *Michel Foucault,* Cambridge, Mass: Harvard University Press

Ferry, L.and Renaut, A., 1990, *French philosophy of the sixties: an essay on anti-humanism,* Amherst: University of Massachusetts Press

Flax, J., 1990, *Thinking fragments: psychoanalysis, feminism and postmodernism in the contemporary West*, Berkeley: University of California Press

Foucault, M., 1970, *The order of things: an archaeology of the human sciences*, London: Tavistock

Foucault, M., 1977, *Discipline and punish: the birth of the prison,* London: Allen Lane

Gallop, J., 1982, *The Daughter's Seduction: feminism and psychoanalysis*, Ithaca: Cornell University Press

Gallop, J., 1985, *Reading Lacan*, Ithaca: Cornell University Press

Granon-Lafont, J., 1985, *La topologie ordinaire de Jacques Lacan*, Paris: Point hors Ligne

Gregory, D., 1994, *Geographical imaginations*, Oxford, England and Cambridge, Mass: Blackwell

Grosz, E., 1990, *Jacques Lacan: a feminist introduction*, London, Routledge

Habermas, J., 1987, *The theory of communicative action. Vol. 2: The critique of functionalist reason*, Cambridge: Polity Press, 1987

Harvey, D., 1989, *The condition of postmodernity: an enquiry into the origins of cultural change*, Oxford and Cambridge, Mass: Blackwell Publishers

Jameson, F., 1977, 'Imaginary and Symbolic in Lacan: Marxism, psychoanalytic criticism and the problem of the subject', *Yale French Studies*, 55/6, 338-395.

Jameson, F., 1981, *The political unconscious*, Ithaca: Cornell University Press

Jameson, F., 1985, 'Architecture and the critique of ideology', in Joan Ockman, Deborah Berke and Mary Mcleod, eds., *Architecture, criticism, ideology*, Princeton: Princeton Architectural Press, 51-87.

Jameson, F., 1991, *Postmodernism or the cultural logic of late capitalism*, Durham: Duke University Press

Jay, M., 1984, *Marxism and totality: adventures of a concept from Lukács to Habermas*, Cambridge: Polity Press

Jay, M., 1993, *Downcast eyes: the denigration of vision in twentieth-century French thought*, Berkeley: University of California Press

Kelly, M., 1982, *Modern French Marxism*, Oxford: Basil Blackwell

Krauss, R., 1993, *The optical unconscious*, Cambridge, Mass: MIT Press

Lacan, J., 1977, *Ecrits: a selection*, New York: Norton and Co.

Lefebvre, H., 1969, *The explosion: Marxism and the French Revolution*, New York: Monthly Review Press

Lefebvre, H., 1976, *The survival of capitalism*, London: Allison and Busby

Lefebvre, H., 1991, *Critique of everyday life*. Vol. 1: *Introduction*, London: Verso

Lefebvre, H., 1991, *The production of space*, Oxford, England and Cambridge, Mass: Blackwell Publishers

Lefebvre, H., 1992, *Éléments de rythmanalyse: introduction à la connaissance des rhythmes*, Paris: Éditions Syllepse

Leupin, A., 1991, 'Voids and knots in knowledge and truth', in Alexandre Leupin, ed., *Lacan and the human sciences*, Lincoln: University of Nebraska Press, 1-23.

Lévi-Strauss, C., 1969, *The elementary structures of kinship*, Boston, Mass: Beacon Press

Lewis, H., 1988, *The politics of surrealism*, New York: Paragon

Lynch, K., 1960, *The image of the city*, Cambridge: MIT Press

Macey, D., 1988, *Lacan in contexts*, London: Verso

Merquior, J.G., 1986, *From Prague to Paris: a critique of structuralist and post-structuralist thought*, London: Verso

Miller, J-A., 1992, 'Michel Foucault and psychoanalysis', in Timothy Armstrong (trans.), *Michel Foucault: Philosopher*, London: Routledge, 58-63.

Morris, M., 1992, 'The man in the mirror: David Harvey's 'Condition' of Postmodernity', *Theory, culture and society* 9, 253-279.

Olalquiaga, C., 1992, *Megalopolis: contemporary cultural sensibilities*, Minneapolis: University of Minnesota Press

Pile, S., 1993, 'Human agency and human geography revisited: a critique of 'new models' of the self', *Transactions of the Institute of British Geographers* 18, 122-139; the quotation is from p. 135.

Popper, K., 1959, *The logic of scientific discovery*, London: Hutchinson

Poster, M., 1975, *Existential Marxism in post-war France: from Sartre to Althusser*, Princeton: Princeton University Press

Preziosi, D., 1988, 'La vi(ll)e en rose: Reading Jameson mapping space', *Strategies* 1, 82-99.

Rajchman, J., 1985, *Michel Foucault: the freedom of philosophy*, New York: Columbia University Press

Rose, G., 1993, *Feminism and geography: the limits of geographical knowledge*, Cambridge: Polity Press

Roth, M., 1988, *Knowing and history: appropriations of Hegel in twentieth-century France*, Ithaca: Cornell University Press

Roudinesco, E., 1990, *Jacques Lacan & Co. A history of psychoanalysis in France, 1925-1985*, Chicago: University of Chicago Press

Silverman, K., 1983, *The subject of semiotics*, New York: Oxford University Press

Silverman, K., 1992, 'The Lacanian phallus', in *Differences* 4, 84-115

Silverman, K., 1992, *Male subjectivity at the margins*, New York: Routledge

Smith, N. and Katz, C., 1993, 'Grounding metaphor: towards a spatialized politics', in Michael Keith and Steve Pile, eds., *Place and the politics of identity*, London: Routledge, 67-83.

Soja, E., 1989, *Postmodern geographies: the reassertion of space in critical social theory*, London: Verso

Derek Gregory
Department of Geography
University of British Columbia
1984 West Hall
Vancouver V6T 1Z2
Canada

3. World time and world space or just hegemonic time and space ?

Milton Santos

Many scholars have devoted perhaps excessive time and talent to discuss the idea of post-modernity. The debate has often centered more on the formal than on the substantive aspects of the question and has therefore only rarely created the conditions for an epistemological renewal of academic disciplines and thus for the understanding of reality.

Among geographers, the debate often repeats the ideas of Paul Virilio (1984), for whom there is no more space and only time exists. What can we do with this metaphor, when the basic material with which we work is exactly the banal space which did not vanish with contemporary acceleration but has merely undergone a quality change ?

We live in an era of paradox; a fact, which has been effectively integrated into the discourse, but is hardly used in the building of epistemological categories, even among those used to work with older dialectics. Today, the same vital impulse (energy) not only generates contradictions within a self-same process, but also creates apparently antagonistic and paradoxical process(es). The truth of this vital impulse is also present among elements that seem to be mutually exclusive. Lacking simple explanations, imagination sometime shrinks. Hence the attraction for metaphor. Yet this should not lead to the death of concepts but, on the contrary, should allow us to emphasize the task of separating metaphor and concepts in order to understand what happens now.

1 World space : world time

Anthony Giddens (1990) recently wrote that we are living in the era of empty time and empty space. We had rather think we live a period of history

45

G. B. Benko and U. Strohmayer (eds.), Geography, History and Social Sciences, 45–49.

which makes it possible to reach a concrete notion of world-time, of full time and full space, and of empirical totality (Santos, 1991).

Let us first lay down the concepts. By *time* we mean the course, the unfolding of events and their weft. By *space* we mean the «milieu», the material basis, or the locus that makes possible events. And by *world* we mean the sum or synthesis of events and loci (places). At each moment, space, time and world are changing altogether. What we have to do is to apprehend and to define the *Present* under this optic.

In his latest book, Regis Debray (1991) makes a parallel between mass media and space, between the role of «mediologues» (media specialists) and geographers. Space is «medium» (media?) in both senses: it is language and also the medium where life is made possible. The perception societies or individuals have of space depends on its historization which is in turn derived from progress in transportation and communication and in the social construction of time. We can find the same idea in a book by Renato Ortiz (1991) whose chapter on space-time is heavily based upon the perception of the changes in the means available to overcome distance by objects (transportation) and by ideas (communications).

Time, space and world are historical realities that must be understood as systems, that is, as mutually convertible in so far as our epistemological concerns are holistic. In any occasion, our starting point is always society in a process of realization. This realization occurs upon a material basis: space and its use; materiality in its various forms, actions and their various modalities.

2 The techniques and empiricization of time

Thus, we turn time empirical by making it material and in so doing assimilate it to space, which does not exist without materiality. Here *technique* comes in as the historical and epistemological link.

On the one hand, techniques make possible the empiricization of time, while on the other they also allow for a clear definition or qualification of the materiality upon which human societies work. This empiricization may thus be the basis for a solitary systemization of the characteristics of each epoch. Throughout history, techniques occur as differently characterized systems. The present technical systems are world-wide, even if their geographical distribution is still unequal, and even if their social use is also still hierarchical. But for the first time in human history, we are dealing with a sole technical system, present everywhere in the West or the East, North and South, which is superimposed upon former technical systems as an hegemonic (technical) system and used by hegemonic economic, political and cultural actors (Santos, 1990). This is an essential point of the globalization process which in turn would be impossible without the unicity.

Consequently, we witness today the emergence of a whole new relation with nature. At the beginning of human history, the unity of nature was reached through telluric forces like, for example, climate, which cannot be understood independently of its world basis (C.A.F. Montiero, 1991; also Staszak, 1994 in this volume). Today, the unifying principle of the world is world society. We thus arrive at the idea of world-world, of a true globalization of the earth, and we do so precisely through this world community which would be impossible without the mentioned unicity of techniques, a unicity, that led to unification of space in global terms and to unification of time in global terms. Space is made unique to the extent that places become globalized. Each place, wherever it be, reflects the world (in what it is but also in what it is not), since all places are potentially intercommunicable.

Wonder of contemporary techniques, all places are united because finally moments are converging. For millenaries, man's history has been a history of diverging moments, a sum of scattered, disparate and disconnected events. For our generation, however, moments are converging; what happens at any place can be immediately communicated to any other place. To this unification at a global scale (M-F. Durand, J. Levy and D. Retaillé, 1992) corresponds time unification. But time is also unified by the generalization of fundamental human needs, tastes and desires, turned common on a world scale (O. Ianni, 1992). If the Universe is defined as set of possibilities, these are worldwide and are theoretically possible at any place, provided that the right conditions are present, Place, or locus is thus the meeting of latent *possibilities* with pre-existing or created *opportunities*. These limit the concretization of occasions.

Science, technology and information make up the technical basis of present social life and in that way they must be part of epistemological constructions aimed at renewing historical disciplines. But we must not forget that our world is extremely hierarchical: We have, on one hand, a new hegemonic technical system, and, on the other hand, a new hegemonic social system with supra-national institutions, multinational enterprises and States at the top, all commanding world-wide objects and world-wide social relations. The result, with respect to space, is the creation of what we call the *technico-scientific milieu* and the emerging of a new system of nature (Santos, 1988).

3 The technico-scientific milieu

Men's environment is no longer what geographers, sociologists and historians used to call, only a few decades ago, the technical milieu. Instead, the technico-scientific milieu is now a geographical environment where territory necessarily includes science, technology and information. In the case of Brazil, practically the whole state of Saõ Paulo - town and country - corresponds to that definition, but also most of South-East Regions and the state of Mato Grosso do sul. We find it in large portions of the states of Goias, Bahia, Mato Grosso and in parts of all other states. In these areas, nothing functions without the participation of science, technology and information, which explains the great importance they hold for intellectual work as a completely new phase (or face) in the process of urbanization.

Science, technology and information belong furthermore to the daily life of the modern countryside: through special seeds, soil care and fertilization, the protection of plants and the existence of a new agricultural calendar. All taken together, this information often makes intermediate cities possess a coefficient of modernity sometimes higher than that of metropoles. We are thus dealing with a new geographical milieu. Mankind has been living for thousands of years in a natural milieu (environment), for a few centuries it lived in a mechanized technical milieu, and now it has arrived in a technico-scientific milieu that is the new face of time and space. It is here that hegemonic activities establish themselves, those with distant relations that participate in international trade.

At the same time, a new system of nature is established (Santos,1992). Primal nature or «natural nature» gave way to artificial nature. Production depends on artifice. The definition of production is no longer intellectual work upon natural nature, but living intellectual work upon dead intellectual work, i.e. artificial nature. This fact was present in cities for a few centuries. Now it is present in the countryside as well. It is a global trend, and places everywhere are defined accordingly.

4 World time or hegemonic time ?

In a world thus remade, we can speak of hegemonic and non-hegemonic times. Hegemonic time corresponds to hegemonic agents and their actions; non-hegemonic time corresponds to non-hegemonic agents and their actions. The concept of hegemonic time presupposes the corresponding concept of hegemonized times. For example, can we speak of a unique time of the city, or a unique time of a region in the same way we speak of a unique universal time ? Groups, institutions and persons live side by side but do not automatically live in the same time. Territory itself is a superposition of differently dated engineering systems which are used today according to different times. Roads and streets are not used in an equal way by everyone. Rhythms - of enterprises, of people - are not the same. It might thus be better to use the term *temporality* instead of *time*.

What we call universal time is that time which embraces all other times, that which gives different values to banal space according to the powers of agents, of the economy, of society, politics and of culture. Hegemonic time, by contrast, is the time of big organizations, of institutions and of states. Between them, we witness a permanent conflict between the hegemonic time of the big organizations and the hegemonic time of the States, as well as the permanently dialectical conflict between the time of hegemonic and non-hegemonic agents. It is in this way, through the use of space and time, that the different facets of daily life are defined.

In the same way hegemonic spaces are created, those spaces which bring forth and embody science, technology and information (i.e. spaces which are loaded with modern rationality), and which become attractive to rational actions driven by global interest. In this manner, we have reached a point in history where the rationalization of society includes the very territory which becomes itself a basic tool of social rationality. Thus it is extremely important to understand how these hegemonic spaces are established in the process of globalisation as places of production and exchange of highest global interests, places where a «world time» prevails and where the forces that regulate actions at other levels are established. It is in this way that diverse times and spaces are united hierarchically in what we may call «world space and time». These are doubtless epistemological realities, but who has already met them in the empirical experience?

We might phrase these observations differently. Today, difference and hierarchy of places are paradoxically a result of their globalization. Idem for times (hegemonic and subordinated times). So-called «world space» is constructed by relations developed between spaces. So-called «world time» is defined by the concretely existing global possibilities and the effectively used global possibilities that are used by hegemonic agents. All other times are subordinated. Here we have thus the empirical basis for a theoretical construction of world space and time; without this theoretical construct it is impossible to understand what is happening today.

References

Debray, D., 1991, *Cours de la Méthode Générale*, Paris : Gallimard.
Durand, M. F., Levy, J. and Retaillé, D, 1992, *Le Monde, Espaces et Systèmes,* Paris : Presses de la Fondation Nationale des Sciences Politiques et Éditions Dalloz.
Giddens, A., 1990, *The Consequences of Modernity,* Cambridge : Polity Press
Ianni, O., 1992, *A Sociedade Global*, Rio de Janeiro : Civilização Brasilera.
Monteiro, C. A. de Figueiredo, 1991, *Clima e Excepcionalismo,* Florianopolis : Edioria da UFSC.
Ortiz, R., 1991, *Cultura e Modernidade*, São Paulo : Brasiliense.
Santos, M., 1988, Réflections sur le rôle de la géographie dans la période technico-scientifique, *Cahiers de Géographie du Québec*, 32, 87, 313-319.
Santos, M., 1990, O periodo técnico-cientifico e os estudos geograficos, *Geografia, Revista do Departameznto de Geografia da USP*, 4, 15-20.
Santos, M., 1991, *Metamorfoses do Espaço Habitado*, São Paulo : Hucitec.
Santos, M., 1992, *A Redescoberta da Natureza*, São Paulo : Universidade de São Paulo, Aula Inaugural da Faculdade de Filosofia, Letras e Ciências Humanas;
Virilio, P., 1984, *L'Espace Critique*, Paris : Christian Bourgois.

Milton Santos
Rua Nazaré Paulista 163, apt. 84
05448-000 São Paulo
Brasilia

4. The language of space

Michel Foucault[†]

Writing over the centuries was governed by time. Real or ficticious narrative was not the only expression of this form of belonging, nor was it the most essential; it is even likely that narrative hid the depth and the law of temporal belonging in the movement which best seemed to demonstrate its qualities. To the extent that writing was liberated from narrative, from its linear order, and from the great syntactical game of sequential tenses, it was thought that the act of writing was relieved of its ancient temporal obedience. In fact, the rigour of time did not affect writing through what was written, but in its very density, in that which made up its singular, immaterial being. Whether directed towards the past, submitting to chronological order or trying to untangle it, writing was caught in a fundamental curve which was that of the Homeric return, but which was at the same time that of the fulfillment of the Jewish prophets. Alexandria, birthplace of us all, had prescribed this circle for every occidental language : to write was to return, to go back to the origin, to relive the first moment; it was to be once more in the morning of time. Whence the mythical function of literature, that prevailed up to our present day; whence, too, its relationship with an ancient past; whence furthermore the privilege granted to analogy and to all the wonders of identity. Whence especially a repetitive structure that denotes the very being of literature.

The twentieth century is perhaps the period when such relationships came untied. The Nietzschean return closed once and for all the curve of the platonic memory, and Joyce reclosed that of the Homeric narrative. Which does not condemn us to the infinity of space as the only other possibility (a possibility which was for too long ignored) but reveals that language is (or has perhaps become) a thing of space. Nor does it matter whether it describes space or merely runs over it. And if space is the most obsessing of metaphors in today's languages, it is not because henceforth it offers the only possible solution; but it is in space that language, right from the start, unfurls, passes over itself,

51

G. B. Benko and U. Strohmayer (eds.), Geography, History and Social Sciences, 51–55.
© 1995 Kluwer Academic Publishers. Printed in the Netherlands.

determines its choices and draws its figures and translations. Space transports language - and in space the very being of language is «metaphorised».

Distance, interval, the in-between, dispersion, fracture, difference are not themes of contemporary literature; rather, the way in which language is given to us today and reaches us is thematized - that which makes that language speaks. Language did not lift these latter dimensions over and above things in order to restore the analogon or the verbal model. On the contrary, these dimensions are common both to things and to language itself : they constitute the blind spot from which things and words come to us the moment they meet. This paradoxical «curve» is different from both the Homeric return or the fulfillment of the Promise - and it is constitutes without doubt the present unthinkable within Literature. In other words : that which makes literature possible in the texts in which we can read it today.

*

La Veille, by Roger Laporte, is closest to this «area» which is both a pale and a formidable one at the same time. It is depicted as an ordeal - as danger and probation, as an opening which creates, but which remains gaping nonetheless, approach and withdrawal simultaneously. It is thus not language as such which imposes its imminence and then immediately turns away. Rather, *«it»* is a neutral subject, a faceless *«it»* through which all language is possible. Writing is only possible if *it* does not withdraw into the absolute of distance; however, writing becomes impossible when it is threatened by the full weight of its own extreme nearness to itself. In this divide fraught with dangers there cannot be (any more than in Hölderlin's Empedocles) either Middle, Law or Measure. Because here the only given things are distance and the vigil of the watchman opening his eyes on a day which has not yet dawned. In a luminious, and totally reserved tone, this *«it»* speaks of the inordinate measurement of the wakeful distance where language speaks. The experience that Laporte talks of as being the past of an ordeal marks the space where its language is given; it is the fold where language increases the empty distance from whence it comes to us and where it separates itself from itself in the approach to this distance over which it and it alone must watch.

In this sense, the work of Laporte, alongside that of Blanchot, thinks the unthought in Literature and draws nearer to its being by the transparency of a language which does not so much seek to rejoin this being as to welcome it.

*

An adamite novel, *Le procès-verbal* by Le Clezio is also a vigil, but in the full light of day. Stretched out diagonally across the sky, Adam Pollo is situated at the point where the faces of time turn back on themselves. Has he perhaps, at the beginning of the novel, escaped from this prison in which he will be incarcerated in the end ? Or else, maybe, has he come from the hospital where he finds the shell of mother-of-pearl, white paint and metal, in the last pages ? And the breathless old woman, who climbs towards him with the entire earth in a halo round her head, is without doubt (in the discourse of madness) the young girl who, at the beginning of the text, climbed right up to her abandoned house. In this fold of time is born an empty space, a distance as yet unnamed where language rushes in. At the top of this distance, a slope, Adam Pollo is like Zarathustra : he comes down to earth, to the sea, to the town. And when he reascends to his lair, it is not the eagle and the serpent, inseparable enemies, the solar circle ,which await him, but the dirty white rat which he tears apart with his knife and leaves to rot on a sun of thorns. Adam Pollo is a prophet in a singular sense: he does not foretell Time but he speaks of this distance which separates him from the world (this world which «came out of his head, so much had he looked at it»). And by the flood of his mad discourse, the world will flow back to him, like a big fish going against the tide, will swallow him and will keep him shut in for an indefinite time, immobile in the barred room of an asylum. Shut up on itself, time spreads itself out now over this checkerboard of bars and of sun;. a mesh, which is perhaps the grid of language.

*

The whole work of Claude Ollier is an investigation of the space common to language and to things; in appearance, it is an exercise in adjusting to complex spheres, countrysides and towns through long and patient sentences which are undone, caught again and locked in the very movements of a look or a step. In actual fact, Ollier's first novel, *La mise en scène,* already revealed a deeper relationship than that of a description or a summary between language and space : inside the circle left blank in an unmapped area, the narrative had created a distinct space, a peopled space that was crisscrossed with happenings. In this space, the one who, in creating described happenings, found himself committed and simultaneously lost because the narrator had had a «double» who, in this same space which did not exist before he did, had been killed by a series of happenings identical to those which were being woven around him; so much so that this space that had never yet been described was only named, related, and surveyed at the expense of a murderous reduplication. Space thus acceded to language through a «stuttering» which abolished time. Space and language were created together in the *Maintien de l'ordre* out of an oscillation between a look which saw itself watched and a look, both obstinate and dumb, which watched it and was surprised in so doing by a perpetual game of retrovision.

Indian Summer (Eté indien) has an octagonal structure. The axis of the absissa is the car which, with the point of its bonnet, splits the scenery in two; it is the stroll on foot, or the drive in a car around town; it is the tramways or the trains. For the vertical of the ordinates there is the climb up the side of the pyramid, the lift in the sky-scraper, the belvedere overlooking the town. And in the space opened by these perpendiculars, everyone of the studied movement unfurls: the turning look, the look which plunges down on the spread of the town as if on a map; the curve of the overhead train which leaps above the bay and then goes back down to the suburbs. Moreover, some of these movements are prolonged, reflected, shifted or fixed in photos, still views, fragments of films.

But all are divided by the eye which follows them, relates them, or itself accompanies them. Because this is not a neutral look; it appears to leave things there where they are but in fact «lifts» them by virtually detaching them from their own density, to make them enter into the composition of a film which does not yet exist and for which the very script is not yet chosen. It is these «views», undecided but under option, which make up the plot of the book with language and between the things which they no longer are and the film not yet existent.

In this new place, anything perceived abandons its own consistence, detaches itself from itself, floats in a space and according to improbable combinations, earns the look which detaches and reties those combinations so well that the look penetrates them, slips into this strange impalpable distance which separates and unites the birthplace of their final screening. Having entered into the plane which brings the look back towards the reality of the film (the producers and authors) as though it had penetrated this thin space, the narrator disappears with it - with the fragile distance built up by his look. The plane falls into a bog which closes over all the things seen in this «lifted» space, which only leaves red flowers «under no gaze» above the perfect and now calm surface, and this text that we read - floating language from a space which has been swallowed up with its demiurge, but which remains present still and for ever in all the words which have no voice left to be pronounced.

*

This is thus the power of language : that which is woven of space, language incites it, gives itself through an originary opening and lifts it only to draw it back into itself. But once more it is bound to space : where else could it float and set down if not in this place which is the page, with its lines and its surface, if not in this «volume» which is the book ? Several times already has Michel Butor set out the laws and paradoxes of this space, which is so visible that language normally covers it without showing it. *Description de San Marco* does not try to restore in language an architectural model which the look can glance at. But it makes systematic and genuine use of all the spaces in language which are closely related to stone edifices : earlier spaces that it restores (sacred texts illustrated by portraits), spaces that are immediately and materially superimposed on painted surfaces (inscriptions and legends), later spaces which analyze and describe the elements of the church (comments on books and guides), neighbouring and correlative spaces which attach themselves rather haphazardly, pinned down by words (remarks of tourists who are looking), near spaceswhose look is nonetheless turned, as if to the other side (fragments of speech). These spaces have their own place of inscription : rolls of manuscripts, the surface of walls, books, magnetic tapes that one cuts with scissors. And this threefold game (the basilica, the verbal spaces, their place of writing) distributes these elements according to a two-tiered system : the direction of the visit (this itself the tangled result of the space of the basilica, of the walk of the stroller and of the movement of his look), and that which is foretold by the great white pages on which Michel Butor has had his text printed, with blocks of words cut out simply through the law of margins, some set out in verses, others in columns. And this organization sends one back perhaps to that other space again which is that of the photograph... There is an immense architecture at the basilica's command, but one which is absolutely different from its space of stones and paintings - rather, it is directed towards it, sticking to it, going through its walls, opening the expanse of words buried in it, carrying back to it every murmur that escapes it or turns away from it, thus causing the games of verbal space to grapple with things to gush forth with a methodical stringency.

«Description» here is not reproduction, but rather decoding : a meticulous undertaking to dislodge this clutter of different languages which things are, to put each one back in its natural place, and to make the book the white location where each one, after de-scription, can again find a universal place of inscription. And there, no doubt, is the being of the book, the object and the place of literature.

Note :

The French version of this text first appeared in *Critique* 203 (1963), pp. 378-382. The present English translation has been prepared by Jenny Money and Ulf Strohmayer.

References

Butor, M., 1963, *Description de San Marco*, Paris: Gallimard
Laporte, R., 1963, *La veille,* Paris: Gallimard
Le Clezio, J. M. G., 1963, *Le procès-verbal*, Paris: Gallimard
Ollier, C., 1963, *Eté indien,* Paris: Minuit

5. Geography before geography :
pre-hellenistic meteors and climates

Jean-François Staszak

Ancient greek geography has been much studied both by hellenists and historians of sciences, especially by those interested in geography. Nevertheless, very few has been written about the pre-hellenistic period, *i.e.* the Aristolelian and pre-Aristotelian geography. Merely a bibliography about the *Historia* written by Herodotus is available, which is at the same time plentiful and very enlightening.

Reasons for this unbalanced historiography can be classified in two fields. On one hand, we can identify the scarcity of original texts that have reached us. Of Hecateus of Miletus, almost nothing remains; the Presocratics' fragments suffer from lots of lacunas (which makes their interpretation very difficult). In return, Herodotus' broad treatise presents a very large *corpus*, which is moreover preserved in good condition. On another hand, it is easy to understand that historians of geography did not feel much of a need to concern themselves with the Presocratics' fragments, Aristotle's texts nor even by the *corpus hippocraticum*.

These do not seem to deal directly with geography and in any case do not pretend to be geographical. Considering their status and contents, we can say that if they are metaphysics, natural philosophy, medecical discourses, they are in no way an organized geographical discourse. However, although these texts are not about geography, it does not follow that there is no geographical knowledge in them. A knowledge, that may indeed constitute their base and/or the referencial frame on which they rely. We are somehow in the prehistory of geography, in a time when geography is not yet «scientifical» and does not even constitute a distinguished field. Here «geography» has not yet demarcated its own discursive field and lacks an epistemological status. However, space and milieu do not only have an objective reality, but are also, somehow, taken in

57

G. B. Benko and U. Strohmayer (eds.), Geography, History and Social Sciences, 57–69.
© 1995 *Kluwer Academic Publishers. Printed in the Netherlands.*

consideration by the people of that time. This paper deals with this geography before geography, this geography of non-geographers.

It presents a double interest. First, the ethnogeographer or the geographer of perception is given here an example of a geographical knowledge, a conception of space and milieu. Secondly, this pre-hellenistic «geography» is of the highest importance as far as the history of geography is concerned. What occurs in this epoch is nothing less than the birth of concepts and epistemological networks that will from hence form the bases of geography for several centuries - and even of nowadays geography, as it is so fundamentally rooted in greek early thought.

It seems that this geographical knowledge is organized in two directions. One developes in the field of natural philosophy or physics, from the Presocratics to Aristotle : this geographical knowledge, built upon the concept of meteors, lays the bases of a physical geography. The other one develops in the field of medicine and - to a certain extent - politics : it centers on the relationship between man and environment and opens the way to a human geography questionning environmental determinism.

1 Natural philosophy, meteors and the prehistory of physical geography

Up to Aristotle (but after him as well), philosophy was not distinguished from what will be later charecterized as «hard science». Philosophers were physicists : they were interested at the same time in metaphysics, ontology and in phenomena, cosmology and nature. Physics (*phusis* means nature) strives to explain everything that exists, especially the way Beings come to be. It is more of a cosmogony than of a cosmology. Among the phenomena that natural philosophers were trying to explain, some play a particulary important part, so much so that these matters in fact seem to monopolize the texts and the debates.

They are the objects that the Greeks gathered in the strange category of the «meteors». This category is so essential that, for a long time, natural philosophy has been called meteorology. Presocratics' fragments as well as their later commentaries by the doxographers show a real obsession for the meteors, which are the subject of an entire treatise by Aristotle, the *Meteorologica* (written *circum* 340 B.C.). But how does this concern geography ?

Greeks' meteors are very different from what we today call meteors. For the Presocratics, meteors are all the ephemeral phenomena that occur in our world : rain, thunder, lightning, but also earthquakes, floods, shooting stars, heavenly bodies... Within the context of magic thought of that time, these strange, hazardous and often threatening phenomena were thought of as gods' signs or manifestations. In fact they were part of many cosmogonical myths. It is here that the passion of natural philosophers is to be found : they intended to offer rational explanations over the accepted mythical ones; in other words, the Presocratics had to take care of the favourite objects of the maguses, deviners and poets.

Basically, the process is not much different within the thought of Aristotle. Nonetheless, Aristotelian meteors differ from Presocratic ones. Aristotle establishes a clear dichotomy between, on the one hand, the supralunar world of the heavenly bodies, made of the fifth essence (ether) and characterized by the perfection of the circular movement, and, on the other, the sublunar world, field of the four elements, disturbed by corruption and linear movement. Thus, Aristotle evacuates heavenly bodies from the field of meteorology, which nevertheless still includes the evolution of shorelines, sources and shooting stars (which he believed took place in the atmosphere).

Behind all this, we can identify a geographical knowledge. In some ways, its content is not all that different from ours. For instance, the hydrologic cycle is clearly identified as early as Thales. Aristotle spread some confusion (according to our point of view) in adding to the humid exhalation (something like a vapour) a dry one, which flows in the wind and burns in the lightning.

Each philosopher built his own theory to explain some enigma like the origin of sea salt or the summer flood of the Nile, which constitutes a rather purple passage within natural philosophy. We do not have the space to present the content of these various and conflicting theories. They co-existed before Aristotle established the first meteorological paradigm, which was to form a base for discussion until the XVIIIth century. Our purpose here is to analyze the exact status of the discourse that formulates these theories.

This discourse is characterized by some epistemological features : the method, mainly deductive, does not really rely on observation (nor on experimentation of course); instead, analogy is used as a heuristic process. We are mainly interested in the discursive status, in the *episteme* that forms its base and context, and in the space it leaves for the development of geographical field.

At that time, the possibilities offered for the constitution of the epistemological field of geography were none. If some theories will be combined with a geographical science thereafter, this does not mean that these theories were geographical. First, because they were not autonomous : they depended on cosmological systems, to which they were a kind of appendix, a deductive development (this is especially true when considering the Presocratics, but still relevant to Aristotle, even if more discreetly). Secondly, these theories ignore several themes which are the concepts at the root of geographical methods.

Aristotelian meteorology does not even mention a geographical space. It does neither approach the question of a spatial distribution of the meteors, nor does try to characterize or to explain the heterogenity of the regions from a meteorological point of view. Aristotle does not regionalize other than on a vertical plane, according to the altitude where meteors are formed and altered. He alomst never mentions the horizontal space of the surface of the earth.

The reasons for this are to be found in the way Aristotle concieves science and space. For him, science could be nothing but nomothetic : science consists of the study of universal causalities. They are perfectly materialized in the movements of the stars, where as the sublunar world is dominated by chaos, hazard and indetermination, as illustrated by the corruption of the bodies, the linear (horizontal) movements, or even the properly monstrous way in which Nature escapes from causality. Consequently, the place where and when a meteor is to occur is completely beyond the reach of prediction; location can not be the subject of any discourse nor, of course, of any science. Location and horizontal movements of the meteors are thus refused a place within science not even within the text.

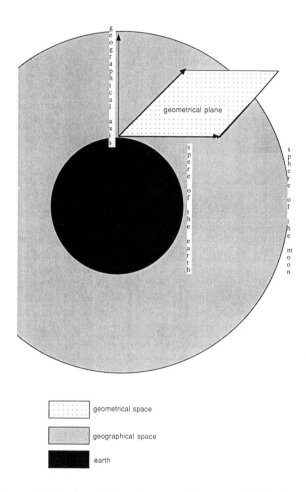

Figure 5.1: Geographical and geometrial space of Aristotle

On the other hand, the vertical directions throughout the «atmosphere», according to which the genesis of meteors is organized, have some characteristics of the perfection of the adjoining supralunar world. This dimension actually is an axis of causalities, in so far as the nature of a meteor depends directly on the altitude where it originates : for instance, Aristotle's dry exhalation, when accumulated behind the vault of heaven, produces the shooting stars; when enclosed in a cloud, it makes thunder and lightning; when flowing back to earth, it brings the winds; when enclosed in the earth, it causes the earthquakes. Vertical location in this stratification can therefore be the subject of a scientific discourse, whereas horizontal location of the meteors at a given altitude (the surface of the earth, for instance), does not present the same opportunity (fig. 1).

This is why Aristotle is not a geographer (fig. 2) - and it is a pity for the development of geography. Of course, it does not follow that Aristotle lived or even thought outside of geographical space. Besides, reading carefully the *Meteorologica* (and some others texts), one can find a climatic zoning of the earth : two polar zones, cold and humid, providing the temperate zones with winds and rains; one equatorial zone, hot and probably dry, from which flow hot winds; and two temperate zones, characterized by the attenuated winds coming both from North and South. There is a climatic pattern, indeed a geographical one (fig. 3).

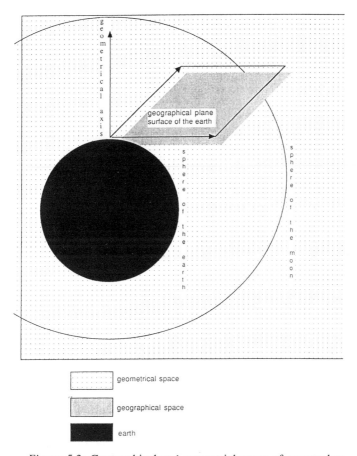

Figure 5.2: Geographical and geometrial space of geographer

Nevertheless, this pattern is never clearly formulated and never constitutes the proper object of any study; it is merely an element of a *Weltanschauung*, which is behind the text and plays an important part in the building of the theories. Indeed, the climatic pattern one could identify in the text is in the mind of the reader (if he is a geographer) rather than in that of Aristotle : this pattern is not properly constructed by the philosopher because, for one, he is not interested in buiding such a theory, and furthermore because he is not concerned by the geographical space. It is therefore not a problem of lack of information. Surely Aristotle had a map at his disposal (it seems that he used it to write the text); he had read some diaries of travels, if no geographical treatises; the Aristotle's *Lyceum* was a center where all sorts of informations were collected.

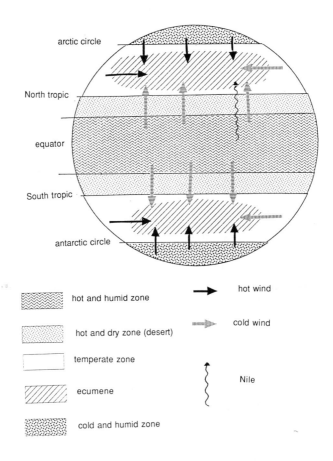

Figure 5.3: The implicite climatic pattern of the earth according to Aristotle

More basically, the philosopher did not intend to develop any geographical research or theory, because his conception of the sciences did not make it possible, did not allow the constitution of an epistemological field for geography. This is why Aristotle has written this treatise about meteors without conceiving a climate. In short, this treatise is nothing but physics.

Still, in some other texts (political or biological ones), Aristotle is indeed a geographer. Not because he writes about space, but because he develops the concept of milieu. In a few but precise words within his *Politeia*, Aristotle is interested in the effect of climate upon societies, particulary in their political constitution. He is more at length where he studies the influence of the environment upon the nature of the animals, and where he explains the diversity and the spatial distribution of animals by the diversity of their surrounding milieus in *Histoire des animaux*.

It is important that the concept of climate is precisely conceived at the moment when Aristotle studies living beings. Besides, it is to be noticed that the place where a living being or a phenomena (a meteor for instance) is located is contingent, and, as such, cannot be the object of any science. On the contrary, the place of a species (that is to say the milieu) is linked to a causal determination, as far as it depends on the nature of the animal : it is hence included in the field of science and is of interest for Aristotle. *This* fish is located in *this* meander - which is the result of chance and there is nothing more to say about it. Yet the fact thet all fish (the species) live in water is not the result of hazard but of causality, and there is a lot more to say about it

However, this topic remains marginal in Aristotle's thought. It is never the exact subject of a study, nor it is the heart of any theoretical system. Furthermore, the question of the milieu is never related to the question of the meteors, there is no physics of the *milieu*. It seems that Aristotle was using somebody else's theory. Indeed, the topic is borrowed - with unequal sucess - from the *corpus hippocraticum*.

2 Hippocratic medicine, *milieu* and geographical determinism

Hippocrates (or a physician of the Hippocratic school) is the one who initiated the debate about the milieu, in his treatise *Airs, waters and places* (written during the second half of the Vth century B.C.), as well as in some other texts of the *corpus*. The text deals with identifying the external causes of diseases, which are to be found in the milieu. As the physician is rather pessimistic, he calls several very different elements to the attention of the reader. Some belong to the nature (*phusis*), like winds, waters and changes of the seasons. They are mainly thought in the framework of two dichotomies, hot *vs* cold and dry *vs* humid. Others belong to human habits (*nomos*) : ways of living (nomadism, sedentariness, alimentation, agriculture, horse riding), customs (ritual deformation of the head by the Longheads), or morals (eagerness to work), or to very accurate practices (archery, babies' swaddling), political organization (mastership, slavery).

Also, Hippocratic theories should not be presented as a «theory of climates», *ie* as a suspect theory of environmental determinism. The human milieu is as important as the physical *milieu*, or even, indeed, more important. Hippocrates notes that Man can escape from environmental pressures by two ways : he can challenge the milieu by modifying its physical characteristics (through agriculture, for instance) and in so doing change the nature of the influence of the environment; he can likewise try to counterbalance the effect of the milieu in his very body (through his way of life and of course medecine, especially through alimentation and diet). Besides, if the physician clearly distinguishes *nomos* from *phusis*, he furthermore shows that they work together and that it is the resulting effect which matters. For instance, the Scythians' lack of fecondity should be explained both by the natural milieu (cold and humid) and by habits (horse riding, wearing of trousers...) (fig. 4).

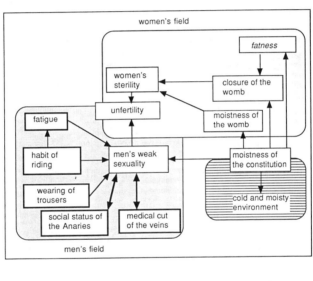

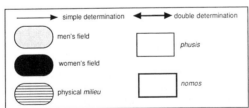

Airs, waters and places, chapters 21 and 22

Figure 5.4: Environmental determination of Scythians' low fecondity, according to Hippocrates

To account for these diseases which can be attributed to the *nomos*, Hippocrates has to consider the diversity of existing *genre de vie* and resulting pathologies. He is thus driven to present a genuine human geography, a geography which reduces variety through the use of typology. Again, we lack the space to elaborate on this point. To compare the position of the physician to that of the natural philosopher, i.e. the meteorologist, we will consider the way in which Hippocrates treats the physical milieu and its diversity, as well as the originality of his approach.

This originality is to be found in Hippocrates' choice of focus. He is primarily looking for pathological influences in the environment. To do so, he has to list different diseases and different environments and, by the method of comparison, achieve correlations. Diversity thus becomes a spatial one; Hippocrates notes that the *milieus* change according to places organized by two axes (fig. 5). The first axis, North-South, polarizes temperature (which is too hot and too cold at the extremes but temperate in the middle); the second axis, Est-West, opposes the region of the rising sun (which is pure, luminous and harmonious) to that of the setting sun (whih is putrid, obscur and unbalanced).

These two axes thus allow the construction of a grid which isolates different cells that correspond to different milieus, in relation to the place they hold in respect to the two primary directions. To each of these cells are matched first, specific diseases and second, specific societies, because the effect of the milieu upon the individual goes hand in hand with its effect upon collectivities. Not surprisingly, the result favours Ionia with its advantagous human milieu (*i.e.* freedom) and its fertile physical *milieu* (*i.e.* temperate climate and contrasted seasons).

Without entering into neither the details of the determining process, nor the geographical values behind it, we can point towards the key role accorded to the concept of constitution, which stands a genuine interface between the human body and its environment. Constitution is, in a way, the nature of man as it is determined by his milieu within a physical framework regulated by the principle of identity. In swampy regions, for example, the main humor is cold and humid : that is to say phlegm, and the constitution is phlegmatical. The diseases here are consequently related to outflow, congestion and overflow : disentery, asthma, catarrhs and suppuration. In hot and dry regions, contrastingly, bile dominates. The bilious are subject to overheating and cracks, and are more likely to suffer from fevers, infection, abscesses, internal laceration and fractures.

This simple scheme is complicated by the fact that, in a given *milieu*, not all the individuals share the same constitution. Hippocrates does not really say why, but we can presume that it is due to former or past migrations, which can bring a hiatus between adapted and unadapted constitution, former and present milieus.

As a result, diseases are ruled by the principle of complementarity between internal milieu (body and constitution) and external milieu (climate and sesonal change). The bilious can compensate the external excess of cold and humidity; the phlegmatic, the external excess of cold and dryness. It is therefore the superpositioning of the same internal and external effect which creates those disharmonies responsable for diseases (which reminds us for the principal of identity). However that may be, normality and healthiness always go with

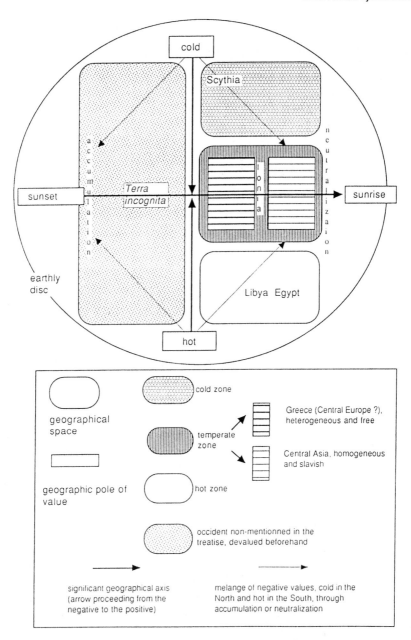

Figure 5.5: Zones, countries and values of Hippocratic space

harmony and regular sucession of the seasons, whereas diseases are anomalies and excesses of any kind, or sesonal disorder.

We have to emphasize that it is the spatial variation of the milieu that is the base of the discourse and the typological tool that permits its edification. If spatial variation, being contingent, was ignored by Aristotle, it is substance matter for Hippocratic theories. This explains why Hippocrates, contrary to Aristotle, constructed a geographical space and thought both milieu and climate. The physician's view here is rather close to that of the geographer, as remains to be seen.

First, the physician is resolutely empirical. Of course, he does not shy away from the formulation of laws but he pretends to draw them directly from experience and observation. Contrarily, presocratic and Aristotelian natural philosophy both are deductive, Aristotle's principled declarations on this matter notwithstanding. Consequently, Hippocrates sets out from the diversity of the phenomenal world, rather than from ideal regularities.

Second, the tradition of natural philosophy, following Aristotelian epistemology, inscribed itself among the theoretical sciences and hence is clearly distinguished from the humanities. On the contrary, Hippocratic medicine precisely intended to link physical *milieu* with the nature of man and societies by the way of constitution. Thinking «the milieu» created great methodological difficulties for the cosmographers even though it was at the heart of Hippocratic system.

Finally, the treatise *Airs, waters and places* was explicitly written for the wandering physicians; its aim was to familiarize the pratictioner with the different diseases he was lickely to encounter on his journey. It is not surprising, therefore, if this text is openly geographical in nature.

Hippocrates was not a geographer. True, he is sometime fascinated by the diversity of the *genres de vie* and of the milieus that inform his discourse, but this latter remains primarily medical. The physician's geographical knowledge has thus nothing exceptional about it : his horizons do not extend very far and the Hippocratic map one could attempt to reconstruct from the text (fig. 6) is not, in fact, all that dissimilar from the Ionian map (we are not far from the parallelogram of Ephorus or the map by Herodotus). The novel and original characters of the text are thus due less to its contents and more to its focus on the milieu : the text constitutes in all likelihood the first systematic treatment of the *nexus* between men and nature. We know today how fruitful this direction would turn out to be, from Montesquieu to contemporary geography.

We would like to conclude by questionning the influence this prehistory of geography exerts on today's geography. If the Aristotelian explanation of the earthquakes and spatialized milieus constructed by Hippocrates are obviulsly far removed from our practices, the epistemological problematics we have mentionned above have indeed a contemporary ring to them.

Of Aristotle's fixation on spatially indifferent physical elments, we have retained what today we call «geosciences» - geology, hydrology... and meteorology. Furthermore, his clear-cut separation between knowledge partaining to the human and the physical sciences still separates our discipline into a «physical» and a «human» geography, a separation that is even more evident in our practices than in our discourses.

On the other hand, Hippocrates opened up a direction quickly exploited by geographers, namely the relationship between men and milieu, which for some has even served as a basis for a definition of the subject. The discredit of geographic deteminism, first launched against Montesquieu's «theory of the climates» and later on against other, more recent, suspect excesses, has probably been an overreaction.

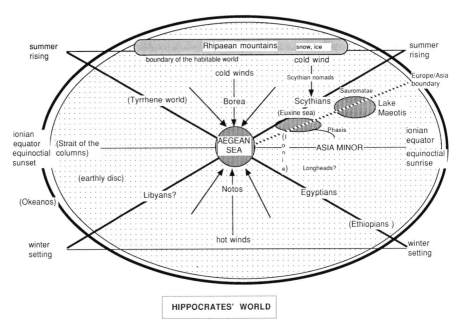

This "map" is based on the text of the treatise *Airs, waters and places*. Directions are those of Hippocrates, but distances are uncertain. All the names are found in the text. Those which can be presumed according to the knowledge of the that time are in parentheses. Names of Longheads and Lybia are followed by a question mark because their location is uncertain.

Figure 5.6: Hippocrates' world

The complete negation of human freedom and the submisssion of men and their societies under natural laws certainly does not produce any satisfying insights. But neither was this the intention of Hippocrates, or, for that matter, Montesquieu. Yet, the violent rejection of determination, if it denies any influence to natural milieus, appears to be based more on ideology than on factual evidence and does, as a result, lead to a regrettable division of our discipline. This division is inforce, even if some climatologists and geomorphologists do indeed attribute heuristic power to the concept of milieu, and pretend not to study physical phenomena but through their importance to human societies (in terms of resources or risks, to give but two examples).

To solve this dilemma, some solutions have been presented by geographers. The majority has accepted the division and rejected the concept of «physical milieu» through focusing geography on space and the study of its structuration. But is not this to throw the baby out with the bathwater, to give in to defeatism? Geographers have probably nothing to win in their abandonment of a problematic but potentially fruitful direction. We have more to gain from Hippocrates than from Aristotle.

References

This paper presents the main theses of my doctoral dissertation : *Météores et climats dans la pensée grecque: les Présocratiques, Hippocrate, Aristote. Essai d'ethnogéographie historique (Meteors and climate in early Greek thought. Presocratics, Hippocrates, Aristotle. An essay in historical ethnogeography)*, directed by Prof. P. CLAVAL, University of Paris-Sorbonne, 1993. The reader will find a complete bibliography in the text, which shall be published.

Aristotle *Meteorologica*, English translation by E. W. Webster (vol. III of *The works of Aristotle* W. D. Ross ed.), Oxford, Clarendon Press, 1931.
Aristotle *Meteorologica*, english translation by H. D. Lee, William Heineman Ltd, London and Cambridge (Mass.), Loeb classical library, 1952, 432 p.
Hippocrates (Corpus) *Airs, waters and places*, english translation by W. H. S. Jones, London, W. Heineman and New York, G. P. Putman's sons, 1923, 157 p.

Jean-François Staszak
Université de Paris IV, 'Espace et Culture'
191, rue Saint-Jaques
75005 Paris
France

6. Geographical systems and the order of reality

Martin Hampl

Increased theoretical developments, interdisciplinary communication and efforts to arrive at a synthesis in post-war geography notwithstanding, the results achieved up until today can hardly be considered satisfactory. Geography has remained a peripheral discipline and geographical knowledge fails to be applied to fundamental scientific concepts (Boulding, 1956, Kedrov, 1961, Teilhard de Chardin, 1956, Zeman, 1985). It seems impossible to find any cognitive principle within geographical reality. Geography, in short, appears destined to continue the praxis of taking up the latest concepts of other disciplines and to submit their results to secondary analyses concerned with the spatial distribution of various phenomena.

Still, the potential possibility of unearthing a genuinely geographical concept or «principle» within reality should be motivation enough to continue searching for it. It is the guiding belief of this paper that the realization of this possibility will depend on the originality of geographical generalizations rather than on the adaptation of geography to the paradigms of more developed disciplines. In this sense the paper at hand is nothing less than consciously immodest and represents an attempt to formulate the general characteristics of geographical systems and subsequently use them for elaborating synthetic concepts of the order of reality.

The debate over the subject of geography was taken as a starting point because the classification problems of real systems and of empirical sciences are closely related to it. Problems concerning the general substance of geographical regularities will be discussed later, at a higher level of analysis. Finally, by developing a simplified specification of the general characteristics into a system of geographical regularities, it will be indicated how possibly to deal with basic epistemological issues relevant to geographical organization. The present paper can only discuss these questions in a limited and simplified way, for more details see Hampl 1988 and 1989.

G. B. Benko and U. Strohmayer (eds.), Geography, History and Social Sciences, 71–80.

As already mentioned, the present point of departure is to identify the subject of geography. This issue is essentially related to the classification problems of real systems and empirical sciences. So far, classifications have been primarily based on the principle of developmental complexity or activity of systems alone. Moreover, the concept of development was conceived in a narrow sense, where only the basic types of elements or elementary forms of the matter movement were distinguished from one another. Within the framework of these essentially linear classifications, there is no room for geography. Evidence of this lacking space can be seen, for instance, in the example of the inconclusive debate between the so-called monists and dualists (Anucin, 1960, Zabelin, 1959, Ljamin, 1978). Therefore, the position of geography within the system of sciences has been either characterized by that of a mere object determination (geography as part of Earth sciences), or else as that of a more or less pragmatic determination (Haggett, 1971, Johnston, 1981).

The surviving ambiguity in identifying the subject of geography can be explained - with exaggerations - by the following extreme yet paradigmatic cases : the subject of geography will be either the spatial structures alone, or the all-embracing content of the geographical environment. In the first case, geography suffers a total qualitative degradation, while in the second, there is an unrealistic augmentation of geography to the position of an omnipotent discipline. In reality, none of these extremes is fully developed in geographical research. But studies in the line of the first paradigm are usually restricted to analyses of spatial variations of different phenomena - often done by non-geographers -, while studies informed by the second paradigm often try to embrace «everything», an aim which can only lead to a naïvely descriptive approach. Geography thus becomes, in the one case, a mere complement to other sciences, or, as in the other, a universal discipline without a fundamental, specific cognitive function.

The solution of the problem can be approached by reconsidering the principle of differentiation and integration of reality. This principle is based on the relationship between the whole and the part, not in a general and relativistic, methodological sense, but in an ontological, and therefore strictly deterministic, manner. The whole is the objective reality for us, or its constitutive «compact» parts, e.g. the Earth's environments, which can be subject to multistep breakdowns. In other words, the principle is reflecting the multilevel character of relationships between elements and their environment, the existence of relatively autonomous, more or less complete, complexes of partial phenomena. The successions of «man - social system - socio-geographical system - complex geographical system» or «organism - biocenosis - bio-geographical system - physical geographical system» can serve as examples.

These successions can be conceptualized in a «cumulative» way, which will lead to a universal concept of reality, an identification of the subject of all sciences (a total concept of the contents of the final complex), or in a «non-cumulative» way, which will enable us to identify relatively autonomous, more or less complete, complexes of phenomena. It is the latter concept that forms the basis for identifying differently complex scientific disciplines which can be characterized as :

(i) oriented mainly towards the inner properties of the individual qualitative types of objects and, consequently, abstracting from their external conditions (disciplines of elementary sciences like biology, chemistry, or psychology etc.);
(ii) oriented essentially towards the external (co-existential, ecological, i.e. complex in our terminology) interactions of qualitatively different phenomena with a consequent abstraction from the inner properties of these phenomena.

Disciplines within the second group represent complex sciences at various levels of complexity. Yet from the point of view of a universe or totality, both the elementary and the complex represent partials, i.e. mere components. By combining the two most essential forms of differentiation in real systems, i.e. the principle of development and the principle of complexity or completeness (complexity as a qualitative heterogeneity and the level of completeness of systems), a basic classification of real systems can be worked out from which a classification of empirical sciences can be deduced. This classification will help to identify geographical systems as well as to find the place of geography within the system of sciences. However, the primary classification of real systems has to be based on the most essential differences among real systems, with regard to their developmental complexity and completeness. For this reason, essential differences among real systems with regard to their qualitative heterogeneity are considered first. Thereafter, the principle of rank, expressing the aspect of size, can also be specified, but this is of secondary importance. Differentiations according to the principle of complexity are expressed by the «element - semi-complex - special complex - final complex» succession (see also the table of classification). As for the principle of development, its point is the succession of «passive - semi-active - active systems», which corresponds to the common differentiation of «anorganic - biological (anorganic and biological, at the level of final complexes) - social (anorganic, biological, and social, at the level of final complexes) systems».

This classification shows clearly that the specific feature of geography is to be found in the highest relative complexity of geographical systems representing special and final complexes at the highest development levels of the organization of reality, starting with the anorganic, or passive, level. Thus the unique and specific feature of geographical cognition lies in the study of external interactions among partial semi-complexes and among partial complexes of qualitatively hybrid phenomena on the one hand, and the whole arrangement of their respective environments on the other (Figure 6.1). The classification of real systems will enable us - by identifying various levels of complexity in real systems - to define the specific features of geographical systems, as well as their position within the order of reality. This is, however, only a first step in the process of identifying the subject of geography. A deeper explanation and a more comprehensive elaboration of such an identification should be connected with the exploration of regularities in the changing organizational set-up of real systems, according to the principle of complexity or completeness, and to the identification of substantial differences between the organizations of relatively

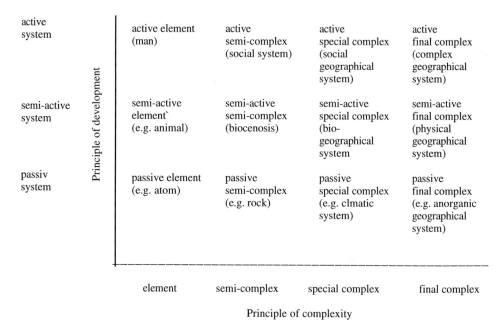

	element	semi-complex	special complex	final complex
active system	active element (man)	active semi-complex (social system)	active special complex (social geographical system)	active final complex (complex geographical system)
semi-active system	semi-active element (e.g. animal)	semi-active semi-complex (biocenosis)	semi-active special complex (bio-geographical system	semi-active final complex (physical geographical system)
passiv system	passive element (e.g. atom)	passive semi-complex (e.g. rock)	passive special complex (e.g. clmatic system)	passive final complex (e.g. anorganic geographical system)

Principle of development

Principle of complexity

Figure 6.1: Primary classification of real systems

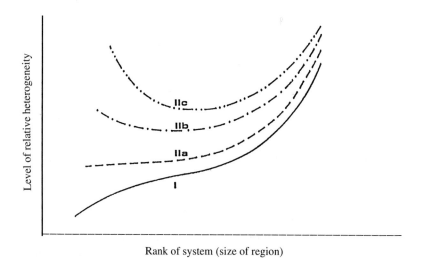

Level of relative heterogeneity

IIc

IIb

IIa

I

Rank of system (size of region)

Figure 6.2: Hierarchy and development of heterogeneity in the distribution of socio-geographic and physical-geographic phenomena

incomplex (i.e. mainly elementary) and relatively complex systems. In this case the key issue is to establish differences in the level and conditions of integrity of individual systems and in the level and properties of differentiation and variety of their generic sets.

Based on various empirical investigations and the generalization derived (see e.g., Korcak, 1941, Hampl, 1971, 1989, Hampl and Pavlik, 1977, Charvat, Hampl and Pavlik, 1978), the following conclusions can be drawn :

(i) There are considerable differences in integrity between elements (atoms, molecules, organisms, people, etc.) and complexes (mainly geographical regions). The elements display a higher level, and mostly internally conditioned, integrity, reflected in their clear-cut delimitation - discontinuity - from the environment, and the integrity of their inner relations. On the other hand, the integrity of complexes is of a relatively low level, and is mostly externally conditioned; their boundaries are not sharply delineated. The definition and delimitation of a plant, for example, is relatively unambiguous, while the definition and delimitation of a bio-geographical region is much more fuzzy.

(ii) An even more important point is that the heterogeneity of generic sets of real systems increases continuously. This process takes place as the complexity or completeness of these systems increases. This is, of course, not related to the definition of the characteristics of relevant generic sets but to other features, especially their size. Thus, on the one hand, there is an analogous «nodal structure» of socio-geographical regions of lower ranks, while on the other, there exists a considerable diversity in their size and importance. In contrast to this, sets of people, plants or atoms of a certain kind are relatively homogeneous, also in respect to their size.

Another important feature is that as the complexity or completeness of systems increases, the diversity of their sets grows in «one specific direction», and in this way a definite - and therefore regular -, the so called hierarchic, type of diversity is realized, involving an inverse relationship between the frequency and the size (or importance) of elements within the set under consideration. E.g. if a group of people is examined in terms of biological, demographical, and potential social characteristics (such as fertility, body height), a relatively high degree of homogeneity will be found within the group, expressed by normal statistical distributions. But if this group is analyzed from socio-economic aspects (income, position within society), i.e. as a semi-complex system, a substantially larger variety will be observed, and the statistical distribution will be shifting towards an asymmetric one. Finally, if considering geographical features (population concentration by settlements, population density), the variety will be exceptionally high and the overall structure of the group will be extremely asymmetric.

The major features of complex systems - which are above all geographical ones - are therefore the following : a relatively low integrity, significant external conditions, an extraordinary diversity in size and, consequently, a considerable hierarchy of their sets. These features differ essentially from those of elements and their sets and this provides a possibility of distinguishing two types of regularities within the reality. Note that so far, regularity has been taken more or less as a mere recurrence of similarity and not as a sort of regular differentiation (see the controversial opinions in geography, Hartshorne, 1959, Bunge, 1962). As for geographical systems, we can also find recurrence of regular hierarchical differentiation of their parts, but this fact already reflects the structural similarities of geographical systems : hierarchical differentiation of partial complex systems indicates the regular differentiation of these systems with regard to size, and also involves a structural differentiation of the corresponding system of a higher rank.

Thus, it can be concluded that there exist two basic types of regularities concerning the conditions and organization of real systems and their sets, mutually combined and inversely related to one another. One is related to the structure of complexes, the other to the structure of elements. To put it in simple terms : reality can be described either as a united resultant final complex (environment) with a developed hierarchical organization, in which partial complexes that vary extremely by size are loosely integrated and highly influenced by the external environment, or as an «atomized» structure, composed of elements with a very strong, mainly internal, integration and a considerable capacity for recurrence (and thereby their generic sets are highly homogeneous). The two corresponding organizational principles are the following :

(i) natural hierarchy which acts dominantly as an ordering principle of the sets of complexes (whose generic sets are therefore characterized by asymmetrical distribution);

(ii) natural capacity for «pure» generic structure, i.e. recurrence in similarity, which functions dominantly as an ordering principle of the sets of elements (the generic sets of which is therefore characterized by normal distribution).

(The non-dominant function of both principles concerns the sets of semi-complexes.) These principles always operate in combination, but their impacts are different and inversely related to one another : the higher the complexity of systems, the more asymmetric is the distribution of their generic sets.

But why does a naturally (ontologically) differentiated hierarchy assert itself as distinct from natural (ontological) generic recurrence (generic structure) and how are they determined ? An answer to this question can be found in an analysis of the general preconditions of real systems for recurrence. This includes an assessment of the general conditions of real systems to reappear, the degree of correspondence between the capacity of recurrence of structural features and that of size.

Essentially, two types of preconditions can be distinguished. The first condition is derived from the probability of recurrence in relation to the density of cases. This condition can be further divided into two forms :

(i) the more numerous a set, the more likely is its capacity for recurrence;

(ii) the simpler, or the less characterized a system, the more likely is its capacity for recurrence.

The dependence on the degree of complexity or completeness will be seen immediately, since its increase implies a decline in density and an increase in the comprehensiveness of systems.

The second precondition of the capacity of real systems for recurrence is a strong integrity of internal relations, or a relationship between internal and external forces in this integration. If internal forces are strong, the realization and creation of similar systems are more likely. Therefore, and in line with the previous analysis, a relatively strong argument can be made for a capacity of recurrence (and natural generic structuration) in the case of elements, but a limited one in the case of complexes.

The above mentioned features can explain the conditions of the natural capacity of real systems for generic structuration (and their various degrees of assertion) but are insufficient to clarify the conditions of their natural capacity for hierarchy. The reason for this is that hierarchical order is not a general, but a specific, type of relative uniqueness (non-recurrence), i.e. diversity.

At present, the hierarchical order of diversity can only be explained in a very general and rather speculative way. To start with, two definitions will be presented, mainly for the purposes of illustration. The first is based on analyzing the possibility of an element to function in a system. It will be assumed that it is more difficult, or less likely, for an element to reach a possible «maximum» (in size, position, etc.) than to reach a «minimum» within a system. This is due to the probability of favorable external conditions as well as competition among elements, and provided that there is an internal order in the system as a whole (like the principle of centrality in location theory). This is a further explanation to the limited possibility of structuration of the whole, see e.g. the constraints in dividing the whole into a large number of small, or a small number of large parts.

This is the starting point for the second definition which derives directly from the possibility of dividing the whole into parts. If all possible breakdowns are given equal chance or taken as equal, it will be found that after appropriate aggregation, the most probable division will be an hierarchical one. The hierarchical division expresses a combination of two extreme cases of dividing the whole into parts, i.e. an equal division of parts and their maximum concentration within one.

It follows from the preceding that although the starting point is the principle of entropy maximization, we arrive at an unusual result. If the starting point were only to give an equal chance to elements to get into equivalent parts, the result would be their even distribution. This indicates a possibility and necessity of two formulations or concepts of the principle of entropy maximization : one from the point of view of elements, the other from the point of view of the whole. The results of these two approaches require different interpretations.

A simple example can serve as an illustration. Let us solve the problem of allocating six inhabitants (n=6) into three regions (r=3). Assuming that no other information is available, we can either start with the even probability of allocation of each inhabitant into each region, or can consider all possible distributions of the set of inhabitants among regions as random. In the first case we get an «even distribution» (2-2-2) as the most probable result, while in the second we arrive at an «uneven distribution», because the highest population density has the least probability. (The task is to aggregate the possible distributions, i.e. 6-0-0, 0-0-6, 2-2-2 etc.) The total number of possible allocations is thus :

$$\binom{n+r-1}{n}$$

where the higher the n and r (within the limits of n>=r>2), the more pronounced is the hierarchical differentiation.

The distinction between natural hierarchization and natural generic structuration, their conditions and differentiated assertions, helps to formulate synthetic ideas about the structural organization of reality and its dialectical substance. On the one hand, we have the whole unique environment as a hierarchically differentiated organization (which is dominantly influenced by external conditions), while on the other hand and within this differentiated environment, we have the details - mainly elements - (organizations which are dominantly influenced by internal conditions), with a relative capacity for recurrence.

This primary structural characterization can be further developed by distinguishing various general levels of generic differentiation (especially elements) and by ranking (especially complex systems). It is however necessary to apply a developmental view first, and particularly to get hold of selective development at the level of both elements and complexes. Due to the limits of the present paper, these complicated analyses cannot be pursued here.

The principle of creating a system of geographical characterizations can also be derived from the generally fixed priorities of hierarchical differentiation, the qualitative heterogeneity and weak integrity of geographical complexes. Such a system should first of all get hold of geographical differentiations which have a multi-dimensional and consistent structure. Furthermore, it should be based on distinguishing certain geographical differentiations, expressed roughly by the succession : «integral hierarchy - set of partial hierarchies - generic differentiation - individual differentiation of geographical systems».

There are several dimensions of this succession. Therefore, the very first thing is to identify these dimensions and their combinations. In a simplified manner again, the succession of dimensions (or of the corresponding principles) can be formulated as follows : basic development differentiation (distinguishing physical- and socio-geographical systems) - rank differentiation (which operates in both systems, determined from the point of view of development) - the principle of complexity (which helps to distinguish between «complex» and «component» geographical systems and operates in a secondary, or limited, way).

According to the above succession, the most important regularities of the internal order of the geographical sphere should be thus connected to the rank differentiation of physical- and socio-geographical systems. Rank differentiation (micro-region - meso-region - macro-region) should be then expressed in a more developed and modified form, as the very hierarchy of hierarchies. This can be done by comparing the internal differentiation of regions with various rank : differentiation within the region with a rank number n+1 is determined by establishing the differences between its constituent regions with rank number n. This latter differentiation (and the corresponding hierarchization) can be established, in a simplified way, by measuring the distribution (even, uneven) of the representative phenomena (such as population in socio-geographical systems or precipitation in physical-geographical systems).

The generalization of empirical studies (see Figure 6.2) enables us to formulate several regularities. First of all, socio-geographical phenomena are less evenly distributed than physical-geographical ones. The most essential is a general difference between physical-geographical and socio-geographical phenomena (or systems) : their relatively uneven distribution has a differentiated impact on rank differentiation. For physical-geographical phenomena, their relatively uneven distribution decreases as the rank of regions decreases, while with socio-geographical phenomena, this process is only valid until the level of meso-region is reached (roughly between 100,000 and 1,000,000 sqkm). From the level of meso-region downwards, the relatively uneven distribution of socio-geographical phenomena increases. These differences between physical- and socio-geographical systems have been increasing with the development process : at an early stage of socio-geographical formation, the two systems were more similar to each other (Figure 6.2).

The gradual lowering of heterogeneity (hierarchization) of physical-geographical, or natural, complexes as their rank (size) decreases, can be explained as a result of the relative determination of a lower rank organization by higher rank ones. Moreover, natural complexes (or complex organizations) are characterized by the prevalence of the impact of external («from above») conditions. It is only with the rise and development of society, that certain, at least secondary, conditions develop which operate in the opposite direction («from below»). This was made possible by a gradual surmounting of natural determination of social activity directed from complex micro-structures towards complex structures of higher rank. In this sense, a socio-geographical «hierarchy of hierarchies» represents a qualitatively higher, active, type of complex environmental organization : besides external natural and internal active conditions which operate «from above» (e.g. the impact of macro-regional socio-geographical differentiation on the development of socio-geographical micro-regions), there is also an activity «from below», though in a secondary and limited way (like the partial surmounting of physical determinations and an active influence of micro-regional socio-geographical organizations on macro-regional socio-geographical organizations).

References

Anucin ,V. A., 1960, *Teoreticeskije problemy geografii (Theorical Problems of Geography)*, Moscow: Geografgiz

Boulding, K. E., 1956, 'General systems theory - the skeleton of science', in *General Systems*, vol.I, 11-17

Bunge ,W., 1962, *Theoritical geography*, Lund: Lund Studies in Geography, Serc. C

Charvat, F., Hampl, M., Pavlik, Z., 1978, *Society and the Co-existancial Structure of Reality*, Prague: Czechoslovak Academy of Sciences

Haggett, P., 1972, *Geography : a modern synthesis*, London: Harper and Row

Hampl, M., 1971, *Teorie komplexity a diferenciace sveta (The Theory of Complexity and Differenciation of the World)*, Prague: Charles University

Hampl, M., 1988, *Teorie strukturalni a vyvojove organizace geografickych systemu : principy a problemy (The theory of structural and development organization of geographical systems : principles and problems)*, Studia geographica 93, Brno: Institut of Geography

Hampl, M., 1989, *Hierarchie reality a studium socialnegeografickych systemu (Hierarchy of reality and the study of social geographical systems)*, Prague: Academia

Hampl, M., Pavlik, Z., 1977, *On the nature of demographic and geodemographical structures*, Prague: A.U.C. Geographica 12

Hartshorne, R., 1959, *Perspective on the nature of geography*, Chicago: Rand McNally

Johnston, R. J., ed. 1981, *The dictionary of human geography*, Oxford: Blackwell

Kedrov, B. M., 1961, *Klassifikacija nauk (The classification of sciences)*, Moskow: Izdatelstvo V.P.S

Korcak, J., 1941, *Prirodni dualita statistickeho rozlozeni (Natural duality of statistical distribution)*, Statisticky obzor 22, Prague

Ljamin, V. S., 1978, *Geografija i obscestvo (Geography and society)*, Moskow: Mysl

Teilhard de Chardin, P., 1956, *Le groupe zoologique humain*, Paris: Albin Michel, Czech translation, Prague: Svoboda

Zabelin, J. M., 1959, *Teorija fiziceskoj geografii (The theory of physical geography)*, Moskow: Geografgiz

Zeman, J., 1985, *Filozofie a prirodovedecke poznani (Philosophy and natural scientific cognition)*, Prague: Academia

Martin Hampl
Department of Social Geography and Regional Development
Charles University
Albertov 6
12843 Prague
Czech Republic

PART III

CITIES AND LANDSCAPES IN TIME

7. LANDSCAPE AS OVERLAPPING NEIGHBOURHOODS

Torsten Hägerstrand

«Landscape» seems to be going out of fashion as a key concept in geography. This is the case also in central Europe, the core area of landscape[1] studies and landscape philosophy. The changing outlook is still more obvious elsewhere. A recent five-year catalogue of doctoral dissertations in geography, submitted to North American universities, only six titles out of about six-hundred include the word landscape. A new *Dictionary of the Environment,* printed in London, does not even list landscape among its title-words, which is very surprising.

The situation was quite different when Carl Sauer began his career in Berkeley by publishing his influential book *The Morphology of Landscape.* At that time the approximately corresponding concept moved on the upward slope of its innovation curve in European geography. As is well known, Carl Sauer himself rather soon found that the concept limited his thinking. Work on the concept of landscape, however, continued to flourish in Europe. In 1939 Carl Troll, the well-known German biogeographer, coined the term *Landschaftsökologie* (landscape ecology). Three decades later - «in the interests of a better international understanding», as he said - he had to give up the term and replace it by the less expressive compound «geoecology» (Troll, 1972: 1).

As a student at the end of the thirties, I was surrounded by the same tradition as the one from which Carl Sauer had received parts of his early inspiration. German thinking permeated teaching in geography in Sweden, but perhaps in a less romantic and more practical Nordic variety. Geography was proclaimed as a unifying branch of learning, containing both physical, biological and human ingredients. The perspective was predominantly historical and landscape was viewed as that object of inquiry which gave the discipline its *raison d'être.* Students had to learn to be at home in the field, in the archives and in the map laboratory. But by the late thirties a couple of manual desk calculators appeared in the department, heralding the dawn of the new numerical era. Such

G. B. Benko and U. Strohmayer (eds.), Geography, History and Social Sciences, 83–96.
© 1995 *Byggforskningsrådet. Kluwer Academic Publishers. Printed in the Netherlands.*

First Published in Gösta Carlestam and Barbro Sollbe (Eds.), Om Tidens Vidd och Tingens Ordning",
Byggforskningsrådet (The Swedish Council for Building Research) T21:91, Stockholm 1991).

was the situation at the time when Richard Hartshorne (1939) argued that the term «landscape» was too ambiguous and not really very essential.

1 Distinctions

When looking back on this period I think it is fair and important to draw a distinction between what those generations of geographers wanted to achieve and the methodological instruments they had at their disposal. The ambition to create a unifying discipline was really quite heroic in an intellectual climate where specialization already stood out as the order of the day. Geography was not the only field suffering from strong tensions between specialization and integration. A rather similar development was taking place in medicine. The art of medicine based on clinical experience which had tried to assist the individual person as a whole, became at an accelerating rate superseded by experimental natural science medicine which only aimed at finding the proximate cause of a disease. Today we know that health care is in trouble, despite the miraculous achievements made by medical technology. Or, as the Swiss historian of medicine Lichtenthaeler (1982: 191) expressed it at a recent Nobel symposium : «Here, a historical 'law' is being verified once again. All great advances are unilateral ; progress is always made at the expense of something».

I believe that one is justified in drawing a parallel between the progress of science and technology on the one hand and man's increasingly destructive role in changing the face of the Earth, including the urban environment, on the other. When saying this I am clearly reflecting my teachers' utopian ideas about the «balanced» landscape and my own understanding of a just and humane society. Despite excursions I have made into the systematic branches of geography, I have over the years maintained the conviction that behind the concept of landscape and its associated historical perspective were deeper aims and values which are probably more important to keep alive today than when they first appeared on the agenda. So many of the major problems of life on Earth today are not isolated from each other but strongly interwoven, and therefore we must learn how to combine separate facts and insights into more all-embracing conceptualizations. This does not mean that we should go back to vague verbal impressionism again. We have to invent more powerful conceptual tools.

My friend and colleague Olavi Granö, geographer at the University of Turku in Finland, has decorated the wall of his office with a remarkable drawing called «The scientific points of view». We see how the scientists eagerly pop their heads each into his own telescope, pointing in various directions and resting upon each other like the beams of a somewhat shaky scaffold. Despite the impressive structure of telescopes, the real world, symbolized by the tulip landscape down below, is out of reach. The drawing was made in the 30s by the Swiss philosopher and geologist C.E. Wegmann while he was engaged in field-work in Finland. He would not have seen the situation differently today. Instead he would have had to add a lot more telescopes. He would probably also have

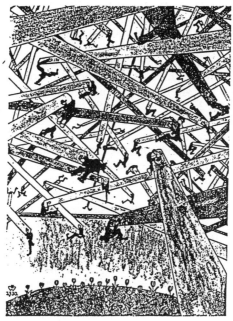

Figure 7.1: Die wissenschaftlichen Gesichtspunkte / The scientific points of view / de vetenskapliga synpunkterna. Drawings by the Swiss philosopher and geologist C.E. Wegmann 1930. Belongs to Professor Olavi Granö, Åbo

had to include some dead tulips. The point Wegmann is making is related to Hannah Arendt's more respectful remark that science has settled for an Archimedian vantage point far away from life on Earth in order to find what is invariant all through the Universe. It does not want to have to do with a «local world».

2 Old movement

It is significant and characteristic for the period that Wegmann included a piece of visual landscape as representing the coherent real world. In this he adhered to a century-old cultural movement in literary, artistic and educational circles in Europe. In fact, geography was rather a late-comer within this movement. Gradually, during the 19th century, landscapes with a more and more realistic content vied with portraits as the dominating subject in painting. Academies of art established professorships in landscape painting, and physicists taught artists the laws of optics (Hildebrandsson, 1897). Some artists were professionally interested in geology and biology. The expanding stream of travel books were supplemented by pictures drawn by accompanying artists. Descriptions and paintings created among members of the educated class a comprehension which made it quite natural for geographers to adopt the term landscape in their professional language.

It is difficult to lift into the open what in some historical periods was seen as self-evident and taken for granted in the conception of reality. Written sources, for obvious reasons, are silent about such matters. One has to judge from circumstantial evidence. J. Ritter (1963: 29) has argued that at the same time as the expanding natural sciences broke up the world into separate objects, the aesthetic realm took over the function of providing a connection between Man and Nature. Oddly enough, the same natural sciences simultaneously made it possible for the human being to feel detached from Nature and take on the role of onlooker.

A fundamental disposition underlying the deep and widespread interest in the concept of landscape as representing Nature during the last century, it seems, was one which viewed everything simultaneously present as inseparable. Even earlier G.E. Lessing, the German dramatist and critic, made some remarkable observations in this direction. In his famous work *Laocoon. An Essay upon the Limits of Painting and Poetry* (1766) he warns poets against empirical descriptions of landscapes, because the consecutive nature of language breaks up into parts what is bodily coexistent. These parts are then difficult or impossible to put together again. What exists side by side becomes transformed into before and after by verbal description (Kortländer, 1977: 28). Painting does not have this kind of limitation. Nor has the map, a geographer would add.

A later indication of how people viewed their surroundings can be found in book illustrations which became more and more common from the middle of the 19th century. At that time the artist usually felt obliged to fill the whole surface inside the given frame. In order to do so he had to invent a large number of details which the author had never mentioned in his story, but which were supposed to have a role in a plausible landscape or interior. Today artists do not feel this obligation. They feel free to simply indicate a situation against an empty background. A corresponding change in scenography is even more striking. The original stage designs of Wagner's operas, for example, are total landscapes full of realistic details, while present-day designs contain only a few symbolic attributes or expose nothing but light on a neutral back-drop.

Is it far-fetched to see this general transformation of outlook from being all-embracing to being selective as a close parallel to the fragmentation of knowledge as illustrated by Wegmann's drawing ?

3 Double Meaning

As the word landscape traveled through the German world of ideas into geography, an older, mediaeval meaning continued to live on : landscape as a smaller territory with internal coherence. To many German-speaking and

Scandinavian writers this double meaning has not posed difficulties. But a fully adequate translation into English seems to be impossible, to judge from Hartshorne's analysis.

The function of the double meaning of landscape is most easily understood in connection with the use of landscape pictures in the teaching of geography in our elementary schools. From the middle of the 19th century the principle was propagated that children should acquire their basic geographical training near home. Then, gradually, more and more distant areas and places should be located on the map and studied. But these other places were also to be comprehended as the homes of other people. The landscape picture then became one of the means of getting across an understanding of life and environment beyond the local realm. Each picture was chosen to show a local landscape (meaning scenery), representing a wider landscape (meaning territory). In this way the two meanings merged. This was also a period when leaders were eager to foster national feelings. Landscape pictures, songs and tales were combined to inspire the children with love for their country. Zachris Topelius, for example, a leading poet and the first university geographer in Finland, published the book *Finland in pictures* in 1845. At the same time he wrote songs, telling us that

Low is the cottage, tall the pine,
loud the roar of the waterfall,
and mighty, more than all human thoughts,
the heartbeat of Nature.

Finally, ideas from landscape painters were adopted by the mass producers of school equipment. Elementary schools were provided with big wall-plates of representative landscapes. These plates helped to form a coherent and common image of the country among ordinary people. I think it is correct to say that patriotic feelings in the Nordic countries became strongly connected with the landscape, something which helps to explain the powerful popular interest in outdoor life, the protection of nature and of places representing a cultural heritage.

«The landscape» as a scientific concept served European geography well for a long time and inspired a huge volume of solid empirical studies. Authors in the first years of this century included only natural elements (Passarge, 1908, Hettner, 1908), but in the following decade human features were also taken into account in the name of the unity of geography (Hassinger, 1917). The fact that the key concept had a vague, common meaning was not felt as a major difficulty. It is not unusual that the direction of research is guided by rather poorly defined images even in the so called exact sciences. But clearly also a theoretical debate developed over the years aiming at greater clarity. According to one leading opinion (Bobek, 1972) the role of landscape geography was to form a scientific link between global systematic geography (*Allgemeine Geographie*) - now largely taken care of by special disciplines - on the one hand and descriptive regional geography (*Länderkunde*) on the other. The study of landscapes aimed at uncovering the united action of the general geofactors under local circumstances.

4 Geomeres

In practical work scholars relied very much on the visual appearance of land-forms, drainage, vegetation and human constructions as the basis for research. Then by observing distributional discontinuities one tried to identify a hierarchy of spatial units, a mosaic, down to elementary pieces called facets or microlandscapes or *geomeres* - to use Hans Carol's (1956) not very widespread term. A major difficulty with the concept of landscape as it is normally understood is this limitation to the visual surface of things. I believe this limitation explains why Carl Sauer with his deep interest in the consequences of human agency through time soon moved beyond his initial formulations. So have most other scholars as soon as they attempted to explain what they observed. In Wolfgang Hartke's and Hans Bobek's social geography the visible landscape as an expression of group consciousness connotes only the basic perspective of their analyses.

One innovative European scholar, however, worked out a different approach which is worth much more attention than it has received so far : Johannes Gabriel Granö , once the leading geographer of Finland, a country very conscious of its landscapes. He argued that the geographic observer of Nature - to which he counted everything bodily - should mobilize all his senses, sight, hearing, smell and touch when, searching for the complex elementary and composite units which make up environment and landscape.

Methodologically speaking, Granö was rather unique, because he not only invented prescriptions; he also used them in empirical work. Granö's approach embraces an anthropocentric natural science (Paasi, 1984). In his *Reine Geographie* (1929) he did not have in mind every man's landscape but that of the trained and disciplined geographer. He was not interested in the private experience. On the contrary, he saw himself as an objective instrument measuring the total character of the environment as far as it could be perceived by the senses. The delimitation of «natural areas» was only a first step. After that, interpretation must follow and about this task Granö wrote :

«An endless chain of causes and effects runs through the forms of the earth's surface, from the soil to the flora and fauna, to settlements, economy, traffic, the development of states, intellectual culture and the history of nations. An almost infinite number of correlations connect different groups of phenomena one to another - directly and indirectly - and thus the idea of an indivisible Whole arises in us» (Paasi, 1984: 7). This sounds rather overwhelming, and it is no wonder that most scientists forge about such complexities and look at smaller and more tractable things in their separate telescopes.

During the 1970s the thinking in terms of sharply delimited spatial units has been under hard attack, predominantly, I think, because of the manner by which one tried to confirm interrelations. The coincidence of spatial boundaries may still be of value for geomorphology and geoecology but this approach has lost much of its power for explaining relations between Society and Nature and inside Society. On the other hand, there is a point in a remark made by a Swiss philosopher (Hoyningen-Huene, 1982: 27) - that it would be an interesting task in itself to investigate the emancipation of the anthroposphere from the other spheres during recent centuries.

A further reason for the declining use of the concept of landscape is more external, as has been pointed out by Gerhard Hard (1969: 262), perhaps its leading critic. Such general ideas expressed by *Ganzheit, Gestalt, Individualität, Zusammanschau,* and other barely translatable words, have lost much of their early substance under the onset of analytical science. Those who still believe that the complexity of the world is worth considering have escaped to the vocabulary of systems thinking. But whatever a landscape as a whole is, I agree with Ernst Neef (1981) that the «compositum geographicum» can only partially be handled in terms of systems.

5 Key concepts

Presently landscape as a key concept seems to have its stronghold in the area it originally emerged from : tourism and a sort of aesthetics. Bookstores offer a rich variety of folio-size volumes, showing in colourplates the face of town and country, but always holding up the impressive or picturesque at the expense of the typical. The purpose is now less to make people love or understand their country than to help them sell it. Recreational and land-use planners and landscape architects demand precise assessments of landscape qualities, an activity also encouraged by land-use legislation in many countries.

To maintain that there is something of fundamental scientific importance *in,* or rather *behind,* the concept of landscape seems to be out of phase with current thinking. Nevertheless, that is what I am going to argue. In his reasoning about what to do with this unsatisfactory concept of landscape, or rather *Landschaft,* Richard Hartshorne (1939: 162) wrote that «the fact that the human mind has a unit impression of a collection of things does not prove for a moment that they have in themselves any relations to each other, other than juxtaposition».

I agree with Harsthorne that the juxtaposition of things per se does not need to make them into something one could consider to be a whole. But I would deny that *juxtaposition* is an insignificant relation. It is rather just the juxtaposition of a mixed assortment of entities - some obviously interdependent and others seemingly only present close to each other - which has called forth a need for a word that communicates these characteristics. To suggest that «landscape» in many cases could be replaced by «area» or «region» is rather analogous to putting a sign of equality between the two concepts music and sound. The first has a definite structure. The second may have one, but this is not implied by the concept.

6 Some speculation

Why the reality which the word landscape tries to denote is so intractable is worth some speculation. First, the local landscape is a practical reality for us in one way or another every minute of our life. But we don't think about it in that way. Most of the innumerable little events are routines and the knowledge needed for handling them is relegated to what Michael Polanyi called the tacit dimension. It would be exhausting if we had to check if there is ground to walk on in front of us for every step we take or to look at a street plan every time we are going to round a corner. We know by heart about up-hill and down-hill and where to go on a sunny day to find grass and shadow. We buy a house with a view and after a few days we are not aware of it unless somebody tries to block it. The combination of features that a person invests in his tacit knowledge is a strictly private set. Each of us appropriates his own little piece from what is there. Yet it is hard for us to fancy what other people's worlds are like. There is much overlap, of course.

In addition, much matter and life is simply present, even close to us beyond the touch of any human being, while to some extent potentially available. In any case, we manage to become so well acquainted with the everyday landscape around us that we forget about its existence at least until something unusual happens.

The second difficulty has to do with the nature of verbal language. It is made for telling stories or expressing feelings and wants, but - as Lessing pointed out - *not* to keep track of what is coexistent. Because of that it takes a lot of effort to think in patterns which are not supported by language. I believe that the strivings in geography to detect such spatially bounded units which were assumed to form real individual wholes were pursued in order to make it possible to classify and name spatial entities, thereby rendering landscape more accessible for treatment by language.

A third difficulty was indicated in Wegmann's drawing. Scientific investigations during the last century have moved in the direction of more and more detailed knowledge about separate functional systems. The ideal has been - and still is I suppose - to find exact causal relationships by controlled experiments. The growth of knowledge within this frame of thinking has been very impressive. It would not have been problematic if it had remained preserved in libraries and the minds of scientists. But scientific findings become applied, usually for the best of intentions. The real world is not a closed laboratory and so applications come to affect the social and the natural worlds in frequently unexpected ways. They encroach upon each other. What seems to be optimal in one perspective comes out as detrimental in another. However, instruments for judging in advance are largely missing.

To recapitulate : our understanding is piecemeal at all levels and it is extremely difficult to communicate across personal experiences, language habits and scientific theories.

7 Common denominator ?

Is there no common denominator of fundamental importance between these realms, between our personal everyday activities, the real events which we tell stories about and the various technical applications of scientific knowledge ? I think there is. They all need a place or places to be, they need *room* at their disposal over a sufficiently long *duration*. And for this duration they need a minimum of «entourage» for support and they must be left in peace by entities and forces which threaten their existence. In other words, some fellow-beings must be *present* and others must be *absent* in the neighbourhood or within reach. These same conditions have to be fulfilled also for natural beings and events in order to allow them to exist and unfold.

Room is not a commodity which is freely available and unlimited. There is always a historically given configuration of rooms for the world's beings to exist in, small chambers for tiny creatures, bigger ones for trees and humans and very big spaces for nations. This given configuration provides the setting for everybody's next step into the future, a step which in its turn cannot avoid changing the configuration to some extent. So it goes on and on for better or worse. Juxtaposition is an important aspect of the probability of outcome. At the now-line new beings are created by reproduction or production. Both forms require room and suitable neighbourhood conditions. Some survive the hour and the day because they manage to cope. Some perish, largely because rooms and neighbourhoods could not be protected from invasion.

Carl Sauer (1974) wrote at the end of his life that «We know that Man is not the master of an unlimited environment, but that his technological intervention in the physical world and its life has become the crisis of his survival and that of its coinhabitants». In order to understand better the nature of those limitations Carl Sauer had in mind, and in order to find ways of holding back, we need not only aggregate measures to compare. We also need improved insights into how the fine-grained budgeting of rooms and neighbourhoods for various beings - men, machines, animals and plants - take place and is affected by human intervention. We are all bound to meet in this budgeting process whether we recognize it or not.

Let me now return to the problem child again, to the concept of landscape. Gerhard Hard has listed eleven different uses of he term in German geography (Hard, 1977: 21). One of these has a venerable history, namely *Erdraum mit seiner gesamten dinglichen Erfüllung,* or approximately «territory with its total filling of things», a formulation which was used by Friedrich Ratzel and perhaps goes even further back. To my mind we have here a definition of the meeting-place I was just talking about. I think that we should continue to wrestle with landscape in this sense until it gives us the answers we need. The crucial question is how to understand the word «thing» in this definition. The practice has been unsettled concerning animals and mobile artefacts. But man himself has usually been considered as a separate agent outside the conceptual scope of landscape, approximately in the same way as the theater makes a distinction between the actors and the stage.

8 «Man landscapes»

Johannes Gabriel Granö in his *Reine Geographie* (1929) takes a different stance. His world is made up of matter, living and lifeless, space and time. He therefore incorporates man himself in the landscape. He speaks about the «pulse beats of life and the rhythm of work» (Granö, 1929: 49). «In certain places and at certain times» he writes (Granö, 1929: 93) «man appears in the landscape in such numbers that one could speak about 'Man landscapes'». But after that he draws a sharp line between body and mind, arguing that «the mental environment is everywhere and particularly in places where the network of settlements and communication is dense, dependent on space in quite other ways than the phenomena of the sensory environment» (Granö, 1929: 178).

Granö was aware of the need to include cultural and social factors in the interpretation of landscape, but to include these factors among the things that seek room and suitable neighbourhoods would have frustrated that initial goal, cherished by geographers at the time, of identifying and naming the units, the wholes, of which landscape was supposed to be made up. If I have understood the text correctly, Carl Sauer in his much less formalistic way saw the human mind represented - by the broad concept of culture - as a force apart from the landscape itself. In this perspective, individual man disappears as an insignificant messenger-boy in between culture and land.

In my own work I prefer to take a different position. The reason for this is conceptually similar to what Clausewitz (1832-34) had in mind when he introduced the concept of «friction» in his theory of war. Friction is the «joint working of countless little circumstances which make real war into something quite divergent from desk strategy».

«One must bear in mind», Clausewitz wrote, «that no part (of any army) is made up of one piece, but that the whole is composed of individuals of which each retains his own friction on all sides». Now, of course, a society is not an army and culture or social leadership are not military commanders, but the actors, after all, are still real individuals and they are also in ordinary civil life exposed to friction on all sides. On the other hand there are also opportunities on all sides. It would be incorrect to overemphasize the negative factors.

A more modern way of expressing both sides of the matter in neutral terms is to say that life is carried on within boundary conditions or constraints. These are partly fixed but partly fluctuating in unforeseeble configurations. Some of these constraints are defined by our intellectual capabilities, some by our social institutions, some - and not the least important - by our bodily existence in physical space together with other people, with animals and plants and the things created by our technology. Juxtaposition is one of the essential determinants of the ways constraints acquire their changing structures. So, after all, the individual is important and if one wants to understand outcomes at the macrolevel one cannot afford to neglect the study of the individual in this environment.

When Granö included human beings in his conception of landscape, he did so because he saw them and heard the sounds of their activities. But then he acted as geographers always have done : he projected his observations onto flat space only. One can read off on his maps the areas where the observer could hear different kinds of sounds at different times of type of the year - but it is all represented as shadings on maps.

9 Space and time together

Now, with some stretch of the imagination one can see with one's inner eye that activities and sounds are not only bound to place. They also have shapes when projected onto time. When space and time are seen together then, suddenly, a new world opens up for investigation. The static map becomes transformed into a plaited weave of trajectories of room-occupying entities which come into being, meet, stay in touch, part and disappear. When the concept of a landscape is extended in this manner, the human trajectories become imbedded among the trajectories of all other beings. And there is no well-founded possibility to single out the human minds from their positions as helmsmen, each preoccupied with his or her share of opportunities and friction. The former observer becomes participant. Sometimes - in order to mark the difference between this image and the classical concept of landscape - I have used the word «diorama», indicating the presence of human beings and their minds and tools in a natural setting. I am not sure this is a good solution : it still accentuates the visual.

There may well be reason to ask why one should continue to pretend that this very abstract and perhaps strange world-picture represents a landscape. Well, the reason is the same as Lessing's, when he criticized poets for tearing Nature into pieces - just like most of science is doing today. It is a matter of principle. I am not talking about statistical aggregates here. I refer to processes bound to individual entities as these are found in their mutual juxtapositions and involved in cooperation and rivalry within a configuration of limited capacity for providing room. All beings on the surface of the Earth - humans and their thoughts, organisms and things - are bound to be able to find room to coexist in close vicinity to each other, or else they become transformed or pushed out into nothingness.

At the moving surface of the present all beings in a bounded area - not only humans - are in a state of uncertainty about the step into the next moment. But the degrees of freedom are limited because of indivisibilities, juxtapositions, speed limits and relative power of beings to join and to repel each other. The mere crowding of phenomena permits only limited access to the future. Therefore, the present on the move is always both a graveyard and a cradle of creation. In Nature some seeds grow if neighbourhood conditions are friendly. Most do not grow. This is all commonplace. But perhaps we do not like to think about the similarities of human intentions and actions. They nearly always encounter general Clausewitz' friction and have to be changed or entirely given up.

Thinking in these terms will help us see that we have been very one-sided in our studies. We have concentrated attention on what has been successful but asked very few questions about what had to give way or be given up in the development process. Thinking in terms of budgeting room and time will help to restore a balance. Today, of course, it is of particular interest to improve our ability to judge the social and environmental costs of technological change. «Investigations of the past and present could be combined with studies of the possible and the desirable» (Zeiher H.J., 1981: 21).

10 Ties

One advantage of adding time to space in a systematic way is that the tie between events in the chosen area and events outside it could be identified and evaluated. What is happening in the human sphere today in any locality is increasingly dependent on actions in far away places. In geography we have been so preoccupied with those horizontal relations - partly because they are easy to map - that we have tended to forget to look inwards towards the places and times where all these interests finally are bound to meet because of the corporeality of the world.

In conclusion, let me demonstrate what the extended understanding of the concept of landscape is aiming at. According to the traditional view, the task of geography was to describe and explain spatial cross-sections of the Earth's surface at chosen points in time. Historical development was most often seen as jumps between cross-sections.

To me this perspective means a tremendous loss of information and insight and it has in fact not been taken too seriously, in particular not by historical geographers in a manner which is not strictly combined with the spatial view. Events still hang somewhat loosely in the air.

Joseph Smith, an American philosopher and musicologist, has said : «One can never ask for time off from life...» In other words, as long as one exists, one is present somewhere, occupying room and involved with beings in one's neighbourhood. The same holds true for animals, plants, artefacts and stones, in the two latter cases if we substitute «life» for «existence».

As a consequence of the impossibility of taking time off, we ought to try to understand what the unbroken persistence of beings from birth to death means to their mutual relations and to the ways things change in the aggregate. The advantage of viewing a landscape in this way is that we discover a fibrosity which binds the past and the future together and reveals relationships which the pure spatial cross-section is unable to capture. We see how neighbourhoods overlap - not only sideways but simultaneously through time.

11 Beyond reach

Clearly, this image is like the classical landscape concept in the sense that a full empirical description is beyond reach. As before we have to be selective in our investigations. There will always be a rest of the world left out, even inside our chosen area. But global understanding can be reached from knowledge of local studies. And when dealing with these, it seems to me that a strict choro-chronology will help us to identify subproblems in improved ways, help us to keep an eye on how we aggregate data and - not the least - help us to offer a design of a meetingplace, where the many actors in politics and business and technology can get a glimpse of the one and only world in which they try to find room for their plans and actions.

I am not sure that Carl Sauer would have liked the rather formalistic approach which I am advocating, but I hope he would have endorsed the purpose.

Note

[1] I use the term «landscape» in the sense of the German *Landschaft* or Scandinavian *landskap*.

References

Bobeck, H., 1972, 'Die Entwicklung der Geographie - Kontinuitet oder Umbruch?' *Mitteilungen d. Österr Geogr. Gesellschaft,* 114

Carol, H., 1956, 'Zur Diskussion um Landschaft und Geographie', *Geographica Helvetica,* 2.

Clausewitz , C. von, 1832, 1980, *Vom Kriege,* Stuttgart

Granö, J.G., 1929, 'Reine Geographie. Eine methodologische Studie beleuchtet mit Beispielen aus Finnland und Estland', *Acta Geographica,* 2, 2

Hard, G., 1969, 'Die Diffusion der 'Idee der Landschaft'. Präliminarien zu einer Geschichte der Landschaftsgeographie', *Erdkunde,* Band XXIII, Heft 4

Hard G., 1977, Zu den Landschaftsbegriffen der Geographie', *Veröffentlichungen des Provinzialinst für westfälische Landes - und Volksforschung des Landschafts-verbandes Westfalen-Lippe,* Reihe 1, Heft 21

Hartshorne, R., 1939, 'The Nature of Geography. A critical survey of Current Thought in the Light of the Past', *Annals of the Association of American Geographers,* Vol. XXIX, 3 and 4

Hassinger, H., 1932, 'Der Staat als Landschaftsgestalter', *Zeitschrift für Geopolitik,* vol. 9

Hettner, A, 1908, 'Methodische Streifzüge', *Geographische Zeitschrift,* 14

Hildebrandsson, H. H., 1897, 'Landskapsmålningar granskade från naturvetenskaplig synpunkt', *Nordisk Tidskrift*

Hoyningen-Huene, P., 1982, 'Zur Konstitution des Gegenstandsbereichs der Geographie bei Hans Carol', *Geographica Hevetica,* 1

Kortländer, B., 1977, 'Die Landschaft in der Literatur des ausgehenden 18. und beginnenden 19. Jahrhunderts', *in* Wallthor, A.H. von and Quirin, H., eds., *Landschaft als interdisziplinäres Forschungsproblem*, Münster

Lichtenthaeler, Ch., 1982, 'Prolegomena to the discussion of the medical session', Nobel Symposium 22, *Science, Technology and Society in the Time of Alfred Nobel*, pp. 186-206, Oxford

Paasi, A., 1984, 'Connections between J.G. Granö's geographical thinking and behavioural and humanistic geography', *Fennia*, 162, 1

Passarge, S., 1908, 'Die natürlichen Landschaften Afrikas', *Petermanns geographische Mitteilungen*

Ritter, J., 1981, *Landschaft. Zur Funktion des Ästhetischen in der Modernen Gesellschaft*, Münster

Sauer, C., 1974, 'The fourth dimension in geography'. *Annals of the Association of American Geographers*, 64, 2

Troll, C., 1972, 'Geoecology and the world-wide differentiation of high-mountain ecosystems,' in *Geoecology of the High-Mountain Regions of Eurasia* (Hrsg. Carl Troll) Wiesbaden

Zeiher, H., 1981, 'Kindheit und Zeit', reprint from Max-Planck-Institut für Bildungsforschung, Berlin

Torsten Hägerstrand
Box 716
22007 Lund
Sweden

8. The urban and the rural :
an historical geographic overview[*]

Clyde Mitchell-Weaver

John Friedmann's 1979 analysis of the contradictions between town and countryside offers a novel conceptualization of urban/rural relations. In my view, however, further reflection is needed as to the ultimate nature of the contradictions between town and countryside if such a reformulation is to have historical validity. In this chapter I will argue that while «urbanness» may reside ultimately in the city's economic functions, it is the political relations between a city and its surrounding region which, historically, have determined their relative equality. Furthermore, it has depended on the type of political organization and the level of technological development as to how this relationship has been defined.

The contemporary physical explosion of urban-like development, and especially the de-urbanization of secondary industry (manufacturing), does not represent a trend towards totalization of the city. Rather, these are facets of the transformation of an «antagonistic historical contradiction» into a new, possibly «non-antagonistic one». As functional economic power strives to divorce itself from its territorial moorings, the cosmic contradiction between territory and function but remains, this may lay the groundwork for workable territorial alliances — a more symbiotic relationship — between the urban and the rural[1].

What is the nature of the city ? What comprises the urban ? How is it different from the rural, and what is the nature of the ties between the two ? Tration.e questions that have been asked many times and I do not pretend to be

---

[*] I would like to thank John Friedmann, Edward Soja, Michael Gunder, Robert Norman, and the late Ernest Weissmann for comments on earlier versions of this manuscript. They didn't necessarily agree with my interpretations, but their advice and criticism were invaluable. This essay is dedicated to the memory of our brilliant late French colleague Philippe Aydalot, whose untimely death detracts greatly from our potential understanding of urban/rural relations.

G. B. Benko and U. Strohmayer (eds.), Geography, History and Social Sciences, 97–132.
© 1995 *Kluwer Academic Publishers. Printed in the Netherlands.*

Period	(A) Point of Diffusion and Examples	(B) Military/Religious Character	(C) Symbolic Stature
(1) Temple/Palace of Occidental Antiquity (3000-1000 BC)	Mesopotamia, Egypt: Sumer, Eridu, Ur, Akkad, Ebla, Thebes, Memphis (Byblos, Damascus, Jerusalem, Jericho)	Fortified temple of the gods, palace-stronghold of their priest-kings	Sacred altar of the god-king; center of all wealth, knowledge and power
(2) Classical City of Mediterranean Civilization (1000BC-500AD)	Greece: Italy, N. Africa, Athens, Corinth, Sparta, Rome, (Massilia, Valentia, Burdigala, Londinium) Carthage	Strong-place of the tribe, public altar of the household gods	Forum of public life, meeting place of the people (locus communis)
(3) Towns and Burgs of the Middle Ages (500-1200AD)	Europe West of the Elbe: Clermont, Reims, Orleans, Maine, Trier, Magdeburg, Regensburg, Canterbury, Nottingham, Chichester	Military fort, fortified cathedral	Place of refuge
(4) Mercantilist City of the Renaissance & Reformation (1200-1700AD)	Western Europe: Venice, Genoa, Florence, Hamburg, Bremen, Antwerp, Bruges, Amsterdam, London, Vienna, Paris, (Boston & Philadelphia)	Focal point of middle class and rural power, birthplace of the reformation and counter-reformation	Hub of commerce, finance and craft industry; center of culture
(5) Imperial City of Industrial Capitalism (1700-1950AD)	Northwestern Europe: (Liverpool, Manchester, Lille) London, Paris, New York, Berlin, Chicago, Moscow, Tokyo	Headquarters of the national guard, imperial barracks, center of evangelical religious sects (Wesleyans, Salvation Army)	Center of industry, capital of the empire and finance, national heartland, focal point of class conflict
(6) Urban Fields of the Post-Industrial Era (1950-)	Northeastern USA: Boswash, Lotharingia, Southern California, Chipitts, Tokyosaka	Military command center, point of dissemination for commercialized "religion," realm of narcistic agnosticism	Seat of corporate power, center of consumption, world core area, information node

Figure 8.1: Transformations of the western city (I)

(D) Predominant Mode of Production	(E) "Non-Essential" Economic Activities	(F) Level of Technological Development	(G) Type of Social Formation
Oriental despotism	Storage, luxury trade, supporting clerical activities and services for urban ruling class	Irrigated agriculture, copper smelting, abstract notation and writing, municipal sanitation	Universal slavery
Classical despotism	Communal market, commerce, craft-industry, supporting services for citizens	Communal and plantation agriculture, bronze and iron metallurgy, road building, aqueducts, beginning of western philosophy and science	"Helotic" or limited slavery and citizen democracy
Feudalism	Supporting clerical activities for military and clergy, vestigial communal market	Extensive domainal agriculture, religious architecture	Seignorial class system (serfs bonded to the land), beginnings of communalism
Mercantile capitalism	Trade and commerce, finance, craft-industry, supporting services for city and hinterland	Development of European high culture, reawakening of science, navigation, guild organization, municipal corporation	Corporate oligarchy and freed men, formation of nation-state and national bourgeoisie, royal absolutism
Industrial capitalism (state "capitalism")	Industry, national and imperial commerce, supporting services, public administration	Industrial revolution, steam technology, electricity, steel, aluminum, the automobile, the airplane, atomic power, economic planning, mass transit	Declining monarchy, bourgeois democracy, (fascism), and authoritarian socialism
Transational "capitalism"	Services, R&D, administration, industry	Aerospace and computer technology: cybernetics, synthetic materials, social planning, bio-technology	Declining bourgeois democracy, welfare, state, emerging functional technology (new totalitarianism)

Figure 8.1: Transformations of the western city (II)

able to provide definitive answers. I do believe, however, that a reasonable approximation of this «essence» of what makes a city different from the surrounding countryside and how the two relate to each other can be found through an historical overview.

Figure 1 presents a chronological breakdown of the transformations of the western city.[2] I have followed the more or less standard periodization of urban development, beginning with the fortified temple/palace of the ancient hydraulic civilizations and ending with the urban fields of late capitalism. While a finer-grained breakdown might have been used, and the dating of period boundaries is certainly open to other interpretations, I believe that the six divisions employed serve adequately for the present discussion[3].

For each period, then, I have attempted to characterize the city and its relations with the countryside according to twelve major historical and geographic traits which seem to dominate most of the literature on the subject. These include : (A) point of diffusion and examples, (B) military/religious character, (C) symbolic stature, (D) predominant mode of production, (E) «non-essential» (or urban) economic activities, (F) level of technological development, (G) type of social formation, (H) physical form and density, (I) clustering and hierarchy, (J) administrative role, (K) dominant town/country ties, and (L) degree of rural/urban equality. No attempt is made to duplicate this information in systematic narrative form; rather, the analysis which follows draws on Table 1 to demonstrate the interplay between four key elements of town/country relations : (1) the role of urbanization in the contradiction between territorial and functional modes of social integration, (2) the city as a focal point of the socio-spatial dialectic, (3) the historical relationship between the city and the State, and (4) the economic role of cities, especially their necessary relationship to secondary industry, commerce, finance and «services».

1 Urban beginnings

According to most writers, the city first appeared in western history between 2500 and 3000 BC in the fertile valleys of the Tigris-Euphrates, Nile and Indus Rivers (e.g., see Childe, 1936 and Mumford, 1938). It appeared as an «implosion» of the traditional neolithic village. This wasn't so much a matter of quantitative change in geographic size or numbers of people : it was a change in qualite. This change — this reflexive involution of agricultural society — took place in the politico-religious realm, based on the con juncture of social and technological development when agricultural techniques had begun to produce a reliable surplus of food. (The same pattern was duplicated in China as well, see Wheatly, 1971, and later happened again in the Western Mediterranean, northwestern Europe, and Meso-America.)

Urbanization, thus, represented a communal «densification» of human contacts and relations, based on production, accumulation and control of a social surplus. All three of these economic innovations, i.e., increased production, deferred consumption, and rationalized distribution, depended on the emergence of a dominant — even tyrannical — locus of political power from within the traditional consultative leadership institutions of clan society. This demanded long ages of internal struggle and, typically, a geographic overlay of different ethnic groups, based on military incursions. It only occurred successfully in a few particular historical-geographic settings, such as the semi-arid river valleys of the Near East, where some form of cooperation and centralized coordination were necessary to ensure a stable social existence (Wittfogel, 1957). Historically, these were achieved through military force, and the resulting political suzerainty required a citadel.

Creation of a central, autonomous political authority took place under different concrete forms, but frequently was associated with the mystical control of water availability or land fertility, and the evolution of «corn god» rituals and an attendant priesthood. From the priestly corporations that formed around such practices arose the State. Its geographic locus, meaning the site of its priestly offerings, grain silos and subsequent fortifications, was the beginnings of the city. And the social order that was thus created was the first to be based on a territorial division of labor (Engels, 1984; Morgan, 1877). As Fustel de Coulanges' studies (1874) of similar developments during the classical period show, destruction of clan society by territorial class society was a slow, unsure process — a process of struggle, overlay and interpenetration. Bookchin (1974) argues that it had reached a kind of dead-end among the Aztecs when the conquistadors first gazed upon magnificent Tenochtitlan. But where it succeeded, it created a new functional ordering to society through territorial integration, brought about by the State and centered on the city.

Four points are of utmost importance here. First, the cosmic contradiction between territory and function took an altogether different form than anything we are familiar with in more recent times. The imposition of «oriental despotism» or the «Asian land system» replaced an almost biological arrangement of functional roles — within the symbol-laden matrix of clan society — by new functional relations based on territorial control; all land belonging to the «higher» community, the State. Territoriality, equated with clan ties by so many writers, was a separate dimension of human consciousness, and it was in aggressive opposition to the functional roles worked out over millennia by the extended family and the tribe.[4]

Closely related to this is the second point. While the city represented the geographic seat of State power and the new social order, it would be a crucial error to equate this with the predominance of truly urban-based economic functions. The material foundations of ancient urban society were agricultural, and the unalterable spatial structure of primitive irrigated agriculture meant that production relations where, in fact, a reflection of rural life. The city was a physical fact. Political power was resident there. And long-distance luxury trade, urban manufactured goods and urban services sprang up to serve the needs of the urban ruling class. But all this — created by the centralized power of the State — was immediately dependent on agriculture, i.e., the countryside. Social processes were evolving in new urban-centered forms, but they were limited, bounded if you will, by spatial structures of earlier human creation.

Thus the third point, as Bookchin observed :

> All cities constitute an antithesis to the land. They are a break in the solidity of agrarian conditions, a germ of negation in the agrarian community. At the same time, however, rural life summons forth the city from its own inner development as a division of labor between crafts and trade on the one hand, and relatively self-sufficient agricultural communities on the other. The emerging city begins by reflecting the social relations in the countryside so that there are different cities more or less corresponding to different forms of agrarian society. In various phases of social development, the city is raised from a distinctly subordinate position to one of equilibrium with the countryside and may remain so for long periods of time or, after overstepping the limits of its rural base, finally yield to the hegemony of the land when the two become clearly incompatible. Viewed over most of precapitalist history, city life did not have as complete an urban basis as it seems to have today. Urban centers were largely the foci of surrounding agrarian relations. They were horticultural clan cities, Asian cities, feudal cities, and even peasant and yeoman cities. Urban life could be clearly understood only by searching back to the economic relationships that prevailed in the agricultural environs. Although city life acquired social forces of its own and often entered into contradiction with the land, the agrarian economy established the historical limits for almost every urban development (Bookchin, 1974: 6.)

We will move on presently to analyze forms of rural/urban equality and urban dominance, but in a very profound sense, during occidental antiquity the countryside ruled over the city.

As a fourth consideration then, where can we find the essence of «urbanism» which separated the city from the earlier agricultural village ? Certainly there was the appearance of territorial society, which was in itself a new manifestation of humankind's relationship to the physical environment. But there was more. The combination of a mystified religious State, with its complex cosmology and theology, explaining its very existence and justifying its privileged position, as well as the need to account for the accumulated wealth which supported urban life, brought into existence abstract notation and writing (see Childe, 1936; Matthiae, 1981; Pettinato, 1981). And this mixture of abstract thought and written records, when unleashed in the dynamic, contact-rich urban environment, produced the beginnings of «high culture» or civilization. According to Mumford :

> What happened... with the rise of cities, was that many functions that had heretofore been scattered and unorganized were brought together within a limited area, and the components of the community were kept in a state of dynamic tension and interaction. In this union, made almost compulsory by the strict enclosure of the city wall, the

already well-established parts of the proto-city — shrine, spring village, market, stronghold — participated in the general enlargement and concentration of numbers, and underwent a structural differentiation that gave them forms recognizable in every subsequent phase of urban culture. The city proved not merely a means of expressing in concrete terms the magnification of sacred and secular power, but in a manner that went far beyond any conscious intention it also enlarged all the dimensions of life. Beginning as a representation of the cosmos, a means of ringing heaven down to earth, the city became a symbol of the possible. Utopia was an integral part of its original constitution, and precisely because it first took form as an ideal projection, it brought into existence realities that might have remained latent for an indefinite time in more soberly governed small communities, pitched to lower expectations and unwilling to make exertions that transcended both their workday habits and their mundane hopes (Mumford, 1961: 32).

Not atypically, Mumford's description may be faulted for romantic enthusiasm, but his central theme seems unquestionably sound.

2 From polis to imperium

At the outset, the classical western city of Period 2, the Mediterranean city of Greece and Roman, shared many of the traits of urban development during the first period. In fact, direct colonization from the Near East certainly played a part in urban origins on both the southern and the northern shores of the Mediterranean. What is important for our analysis here however, is the subsequent train of events. For reasons not dissimilar perhaps to the logic of later developments in central and western Europe, the centralized State of proto-typical oriental despotism was unable to establish itself firmly in Attica or the Peloponnese. Crete provided an island example of the totalizing State, as did the transient reign of various early Hellenic kings and the Etruscan rulers of Latium. And to be sure, new forms of urban-focused centralism arose later with establishment of the Macedonian Empire (338-323BC) and the Roman praetorium (31BC-476AD), but in the interim the universal slavery typical of the Asian land system gave way to an era of «Helotic» or limited slavery and citizen democracy. Before the city became a tool for spreading Alexandrian and Roman imperialism, it was the center of a restricted form of yeoman democracy.

I shall not dwell on the wonders of early Mediterranean democracy or the accomplishments of the Greek city-states. Marx (1857a, 1857b), Fustel de Coulanges (1874), Mumford (l961), Bookchin (1974), and Braudel, ed. (1985) provide analyses of their economic basis, and religious and political culture. The general intellectual and high-cultural achievements of this Mediterranean world are, of course, widely chronicled, and generally accepted as one of the foundations for all later western civilization. My purpose here is only to underline the essential changes in urban life and their implications for rural/urban relations.

In Greece and Italy the economic and technological imperatives did not exist to provide infrastructural support for the extended territorial State. Religion remained centered on the gens — the tribal clan — and the god Terminus presided over family boundaries which divided the communal lands. Clan elders used this persisting rural power base to resist and temporarily contain the pretensions of urban monarchy. Feudal economic relations in the time of Odysseus were gradually transformed into an oligarchy of yeoman farming families. Cities such as Athens became the vital center of public affairs and political administration, as well as a collective alter to the household gods, but few hard distinctions separated the free city dweller from freemen on outlying farms : both were citizens of the polis. The city served as a market-place and public forum in the agora, and a temple and military citadel in the acropolis. Broader trade and commerce expanded, and urban services and civic life flourished. The popular assembly in Athens, though ultimately more democratic and participatory than the representative Roman Senate, formed a unity between rural clan-based power and the rulers of the city. An accommodation was made between the ethnic order of the clans and territorial principles of political economy. A balance was struck between town and countryside, which were intermingled both economically and politically. (In Rome, this unity was eventually destroyed by the rise of a separate patrician senatorial class and displacement of the rural gens by massive slaveholding estates, the latifundia; see Plutarch, 1958.)

These social arrangements, though, were still contextuated by the production relations and spatial structures of relatively primitive agricultural technology. Agriculture continued to provide the economic basis of the community, and actual agricultural production was organized along the lines of a modified «classical» despotism. Social formations of this type were characterized by «Helotic» slavery and a form of limited, ruling-class democracy, which penetrated from the countryside to the very core of urban life. Most all «non-essential» or urban economic activities were below the dignity of citizenship, so commerce, finance, craft-industry and urban services were all put in the hands of an exploited urban slave class; as was urban agriculture, a relatively important aspect of early Mediterranean city life. Only participatory government, i.e. the State, and intellectual pursuits of the less focused sort — philosophy, science, literature — what might today be referred to (a little crassly perhaps) as quaternary services, remained in the hands of the citizenry .

The truly artificial nature of city life under the classical dispensation only became apparent when the «natural» boundaries of the city-region were breached, and the social processes and structure of the polis were extended spatially to encompass the bounds of empire. The city indeed became the political master of the countryside, although citizenship, even in imperial Rome, remained open to anyone in the empire who was legally free to embrace it. The city grew to gigantic size — Rome becoming the first millionaire metropolis. But this was entirely parasitic growth : urban economic functions could not break themselves free of territorial control :

Once trade and the free cities acquired cosmopolitan proportions, two alternatives confronted the ancient world : either mercantile relations would expand to a point that would produce an authentic capitalist economy or the cities would become parasitic entities, living in vampire fashion on the agricultural wealth of the older social system in the Near East and North Africa. The realization of the first alternative was almost completely precluded by the nature of Mediterranean economic life. Trade, while growing considerably, could never reach sufficient proportions to transform Mediterranean society as a whole. There was simply not enough quantity. Although commerce managed to undermine the small-holding, which gave way to large-scale agriculture in Latium, the free cities were too few in number and much too weak economically to dissolve the self-contained wealthy land systems of the Near East and open them as commercial markets. The Asian land system imposed the same limits on the development of capitalist production abroad which confronted its domestic commercial strata at home. Owing to its solidity, it closed off the only potential market of sufficient dimensions that might have transformed mercantile capitalism into industrial capitalism. Ancient trade remained primarily a carrying trade, a cement between the free cities and economically impenetrable societies based on time-honored agricultural ways.

The «Fall of Rome» can be explained by the rise of Rome. The Latin city was carried to imperial heights not by the resources of its rural environs, but by spoils acquired from the systematic looting of the Near East, Egypt, and North Africa... Having passed beyond its domestic limits, Rome «fell»' in the sense that the city contracted to its own agrarian base — and declined even more as a result of the enormous urban heights from which it had fallen... Lacking an adequate agrarian and industrial basis of its own, Rome had swollen to enormous dimensions around a system of plunder and parasitism. The city had turned upon the land and introduced inefficient — even destructive — forms of agricultural exploitation, such as slave-worked latifundia owned by absentee proprietors. Not surprisingly, Rome succumbed to these internal weaknesses when the parasitic system overreached itself and began to acquire less than it lost... The «fall of Rome» as a city, quite aside from the destiny of the empire, was a local «retrogression»... With the rise of feudal society, the European continent was thrown back upon its own mainsprings. A new relationship between land and city began to emerge, one that initiated an authentic development toward more advanced social relationships (Bookchin, 1974: 22-5).

3 Medieval towns and the rise of the middle class

The final occupation of Rome by Odoacer in 476AD did not, of course, mark the immediate end of the imperial Roman network of roads and colonial cities.(5) The spatial structure of imperial rule lingered on more or less in tact for the best part of two centuries, only gradually succumbing to the increasing political and social anarchy of Merovingian Europe (see Pirenne, 1958; Bloch, 1961). Even then (and yet today, for that matter), the small, isolated towns and burgs of the early middle ages remained imbedded, for the most part, in a geographical matrix laid down by the Roman imperium.

In describing the role of towns in the rise of medieval feudalism and the subsequent transition to mercantile capitalism, it will be useful to discuss Periods 3 and 4 on Table 1 together. By the time Charlemagne came to power in Aix-La-Chapelle (768), Europe had recovered from the worst chaos of the «Dark Ages;» the Muslims had been defeated by Charles Martel at Tours and the social relations of a new territorial political-economic order were well entrenched. The Carolingian Empire, encompassing all of western Europe from Rome and the Pyrenees to the southern borders of Jutland, and running from Rennes in the west to Vienna in the east, was, however, a «monumental piece of archaism». It had no substantial material basis and was even dismantled symbolically shortly after Charlemagne's death by the Treaty of Verdun in 843. Centralized political power, which had been an urban attribute in one form or other throughout most of Periods 1 and 2, was now almost totally undermined by a fragmented territorial economic order. The city, as such, disappeared for all practical purposes for a period of perhaps one hundred and fifty years.

To understand this new turn of events, however, we must look back again to the dissolution of the Roman Empire. Bookchin may have been right when he argued that, «The 'fall of Rome' as a city, quite aside from the destiny of the empire, was a local 'retrogression,'» and that, «the European continent was thrown back upon its own mainsprings». But an overly economistic interpretation of Rome's final collapse — a narrow focus on the fact that the «parasitic system over-reached itself» — sheds too little light on the matter. Even conceding the imperial capital's increasing extravagance and gluttony, it must be remembered that the empire had substantially the same boundaries in 400 that it had during the reign of Hadrian (117-138) (Palmer, 1961). (Albeit that after 305 much more of the wealth of the East must have stayed in Byzantium.) The slave-estate system of agricultural «mining» had been in existence since before the fall of the Republic (31 BC), being one of the very causes of the destruction of the Roman gens and the final alienation of the patrician ruling class. And surely the depravity of the emperors who came after Theodosius was matched in equal measure by Augustus' immediate successors : Tiberius, Caligula and Nero![6] I submit that breakup of the empire and the economic decline of the city itself was rather more complicated than Bookchin's explanation would suggest, and that its central causes were — at least immediate temporal causes — military and political.

Gibbon (1788), in his monumental history of Roman decadence, put forward several intertwined reasons for the empire's slow decline : (1) the irresponsibility of militarization, (2) the self-betrayal of barbarization, and (3) the dissipation of energy by the sectarians. Leaving aside Gibbon's acerbic indictment of Christianity, there can be little question that transalpine Europe was held to the mammilla lupi by force of arms, and that «barbarization» of the legions, started under Diocletian and completed by Constantine, sealed the empire's fate, when the Huns pushed the Goths, Visigoths and Vandals southwards toward Italy and the imperial city. Rome's own barbarian generals deposed the child-emperor Romulus-Augustulus in Ravenna (476), and received the Ostrogothic kings without opposition.

The functional economic ties which held together Rome's territorial empire had been imposed by the political power of the Roman State, acting through its armies. When Roman arms, and hence the Roman State, fell to defeat, or better yet, defected, the economic system which had supported it also collapsed. Commerce could no longer move safely along the imperial highways, and the economic raison d'etre of the latifundia vanished. What is more, ravaged civil society proved increasingly unable to maintain even a minimum of social peace. With the continuing influx of Germanic invaders, the scattering of tens of thousands of slaves, and complete destruction of the colonial urban economy, primitive chaos reigned in place of the Caesars. In order merely to survive, colonials, slaves and Germans alike were forced to adopt a pan-historic system of communal agriculture.

And given the total barbarity of the times, freemen — yeoman farmers of whatever origin — had to huddle close together just to preserve life and sustenance. Among the Franks and Germans, agricultural communes became layered by successive waves of conquerors and the conquered. The more vicious and astute military captains conquered most, and also provided the best chance of protection and survival to their «neighbors». In such a situation, the number of «barons» (i.e. freemen) decreased rapidly. To hold off rapine from afar, tyranny and domination from within was the levy. Needless to say, here we have arrived at the beginnings of a new form of political order, which eventually established a new form of territorial society, based on new forms of economic and social relations : feudalism.

Feudalism, of course, wasn't really new; I have already mentioned its temporary role in the founding of the Greek city-states. And although the outcomes of medieval development were superficially similar to the classical polis by the beginning of Period 4 (see, Griffeth and Thomas, eds., 1981), the historical process was entirely different. For in this instance, agricultural relations were left outside the city walls. The relatively balanced city-regions of the High Middle Ages, and the carefully tuned social formations of the early mercantilist city, were truly urban in their origins. As I will argue, on political grounds, the city first established its equality with the countryside and then came to dominate it. By the end of Period 4 the city had contributed significantly to resurrection of the centralized State, and lain a new functional system of economic relations over the spatial structures of territorial feudalism.

Despite their extreme interest, analysis of rural feudal production relations and social forms would take us too far afield in the present context. I will limit the discussion, therefore, to non-agricultural settlements.

The towns and burgs of the early middle ages were by no means urban places, although the towns often occupied the physical shell of former Roman colonial cities (Pirenne, 1925, Chap. 3; 1958: 197-209). Their primary functions were ecclesiastical and military, respectively; their real role in medieval society being protection in times of extreme emergency. Although there were a few larger trading centers, such as Bordeaux, Rouen and Augsburg, commerce had ceased in any significant sense. The local periodic markets were a mere vestige of earlier institutions, serving the needs of the «urban» clergy, military and their servants. Rudimentary urban services and crude craft activities served the same ends. But there were almost no broader ties between town and countryside, even political administration had fled to the rural strongholds of the feudality. There was no general urban population in the towns, and the countryside organized economic production as well as political life in nearly self-sufficient demesnes. In the 9th century towns were simple, dependent consumers; and many of the foodstuffs which they consumed were even produced within their own ramparts.[7] There has apparently never been a more hermetically sealed human environment. Functional geographic ties were reduced to an absolute minimum and small-scale territorial bonds reigned supreme.[8]

According to Marx and Engels (1846) :

> ... the chief form of property during the feudal epoch consisted on the one hand of landed property with serf labor chained to it, and on the other of the labor of the individual with small capital commanding the labor of journeymen. The organization of both was determined by the restricted conditions of production — the small-scale and primitive cultivation of the land, and the craft type of industry. There was little division of labor in the heyday of feudalism. Each country bore in itself the anti-thesis of town and country; the division into estates was certainly strongly marked; but apart from the differentiation of princes, nobility, clergy and peasants in the country, and masters, journeymen, apprentices and soon also the rabble of casual laborers in the towns, no division of importance took place. In agriculture it was rendered difficult by the strip-system, beside which the cottage industry of the peasants themselves emerged. In industry there was no division of labor at all in the individual trades themselves, and very little between them. The separation of industry and commerce was found already in existence in older towns : in the newer it only developed later, when the towns entered into mutual relations.
>
> The grouping of larger territories into feudal kingdoms was a necessity for the landed nobility as for the towns. The organization of the ruling class, the nobility, had, therefore, everywhere a monarch at its head (p. 46).

    This latter contention, i.e. that fragmented territorial society, as well as the towns, required the expansion of feudal kingdoms leading eventually to the nation-state — has proven a thorny question for later writers. How did the centralized State reassert itself ? What was its relationship to the landed aristocracy ? And why did its re-emergence coincide with the rise of mercantile capitalism and appearance of the bourgeois city ? By the 13th century Europe had witnessed the rise and fall of four maritime «empires» : the Italian city-states, Venice, Genoa and Pisa; the Lombard League; several generations of Hanseatic trading confederations, centered on the Baltic and North Seas; and the famous textile cities of Flanders. The close of the 1500s saw unification of the first modern nation-states : France (1480), England (1485) and Spain (1492). But what were the causes of these transformations of feudal society ? What was the functional importance of cities ? How did relations between city and countryside change ? What were the economic and political bases of the new nation-states ? The rise of medieval communes and their later transformations, first to bourgeois trading cities, and then into national industrial centers is one of the most widely studied phenomena in human history. And rightly so, because this increasing subordination of territorial to functional interests — the rise of capitalism — marked the birth pangs and early adolescence of today's global society. I can only attempt to sketch its broadest outlines.

    The communal movement became an important force in feudal Europe during the 11th and 12th centuries (Pirenne, 1925: 69). There is significant contention as to how this momentous trend began, but Bookchin captures the essential features :

> Fortunately... for the cities, European feudalism remained at chronic war with itself. This not only promoted further decentralization but often provided urban communities with a wide latitude for independent growth. By the tenth century, the mutual pitting of French baronies against each other had divided the country into some ten thousand political units. When European cities began to emerge, they found an agrarian society incomparably less unified and materially weaker than the domineering and wealthy Asian land systems of the Near East and North Africa. Given time and the steady settling of the continent, many medieval cities freed themselves from the control of the feudal lords and achieved a modest dominance over agrarian interests (Bookchin, 1974: 37).

    Bookchin, thus, identifies two primary factors : (1) the segmented nature of feudal political suzerainty and the resultant warring, in which each barony developed its own marginal trading center, and (2) the competitive settlement of new land, which often required particular concessions and modifications of feudal relations to attract settlers : «new towns» in which feudal obligations such as corvee labor were replaced by in-kind and even cash payments, and municipal «freedoms» were guaranteed by legally binding writs and charters. (Merrington, 1975, in his interesting treatment of the topic, reinforces Bookchin's generalizations.) Another related factor, underscored in the famous Dobb/Sweezy debate over the transition from feudalism, is (3) the fact that population growth in rural feudal society, after the establishment of relative peace and prosperity, could not be supported by existing agricultural production relations, land availability and technology. Some of this increase had to be absorbed within an alternative structure, if it was to survive. Finally, Bookchin adds another point, (4) :

> ... the medieval commune was devoted almost entirely to handicrafts and local trade. The towns of the high Middle Ages were primarily marketplaces and centers for the production of commodities... . For the most part... medieval communes furnished the skills and products which could not be acquired from the manorial domestic economy. Thus towns never suffered from any confusion about their functions or about the factors which determined their destiny. They had a reasonably clear self-understanding of their commercial and craft interests. Far from being distorted like their antecedents into pliant instruments of agrarian classes, they jealously guarded their autonomy and provided a hospitable environment for independent traders and handicraft workers — the precursors of the modern bourgeoisie.
>
> Yet the medieval commune was a feudal, not a bourgeois, city. *Essentially, its economy was based on simple commodity production — a mode of production in which craftsmen use the marketplace to satisfy their needs, not to accumulate capital.* Although goods were produced for exchange, that is, as commodities (to use Marx's conception of the term), the owner of the means of production remained the direct producer rather than a bourgeois 'supervisor' of productive activity. To be sure, a master craftsman was aided by apprentices, but the latter could realistically aspire to become master craftsmen in their own right once they acquired the skills to do so. In typical feudal fashion, guilds regulated economic activity down to almost the smallest detail; the output, quality, and prices of goods that found their way to the marketplace were carefully supervised by craft associations of master workmen...
>
> In so self-contained and self-fulfilling a society, then, how did it come to pass that these simple commodity relations were supplanted by bourgeois ones and the beauty of the medieval commune by the blight of the bourgeois city ? (Bookchin, 1974: 38-9, emphasis added).

This question brings us to the very heart of the debate over whether the rise of cities was an «internal» feature of feudal society or, rather, «external» to the feudal mode of production.

The medieval commune was, perhaps, the first truly urban creation in history (Bookchin, 1974: 30, Merrington, 1975: 70-77).[9] While it has been demeaned by Marx and Engels as merely another focus of exploitative feudal relations, it has been equally celebrated by other writers, such as Weber, Pirenne, Mumford and Bookchin. In some ways the commune resembled the classical polis; in its highly structured participatory order, in its sense of civic pride and «belongingness». After a fashion, it too was open to citizens (or at least residents) from the countryside — if they could only escape to its sheltering walls and breath its «free» air. Its urban economic functions, seen from one perspective, grew up to fill a void left by the fragmented territorial organization of the manor economy. For some period of time at least, individual towns stood in relatively autonomous relation to the surrounding countryside; a tenuous balance had been struck. This situation did not last, though. Autonomous corporate development gave way to the formation of urban leagues, and, eventually, genuine urban dominance. Why ?

Among economists, speculation about the role of cities in economic history began with Adam Smith (1776); the earlier French «physiocrats», by virtue of their belief in the unique agricultural origins of wealth, took little or no interest in the question. (They were, in fact, conservative supporters of the ancien regime.) Smith's well known argument was that towns acted as the nexus of long-distance trade — as commercial entrepots — which set the whole wheel of capitalist accumulation and development in motion. Subsequent developments : the qualitative transformation of intra-regional trade, decline of the guild system and rural feudal relations, the ascendance of merchant over landed lord, and, finally, the rise of industry. All this, i.e. the wealth of nations, was a product of the city (see Mumy, 1978).

Marx's analysis (1847, 1857a, 1857b, 1875; Marx and Engels, 1846) emphasized the importance of internal production relations for the decline of feudalism, but he too assigned an extremely important role to merchant capital — the «first free form of capital» — in the rise of its successor, capitalism. And the next influential commentators, Max Weber (1905) and Henri Pirenne (1925), re-emphasized the paramount influence of external trade relations on formation of the bourgeoisie, and centered this transformation around the medieval city of northern Italy and Flanders; especially Flanders, where the wool trade lead to true entrepreneurial competition, industrialism, and urban dominance. Two decades later, Dobb (1946) tried to redirect attention back to internal class relations, but was severely criticized by fellow Marxist Paul Sweezy (1950). The ensuing debate, led off by Sweezy et al. (1954), provided little final resolution of the problem, leaving it to Bookchin (1974) and Merrington (1975) to attempt a resolution. Both these latter writers (a libertarian and a Marxist, respectively) emphasized : (1) the crucial political autonomy of towns, within the system of parcellized feudal territorial relations; (2) the subsequent transformation of urban commodity production, which later became generalized throughout surrounding areas; creating (3) abstract production, abstract labor and abstract capital (i.e. «the market»); as well as (4) windfall middleman profits, extracted from functional «city-regions» and, eventually, national markets. The spatial structure of feudal relations (primarily political, but also economic) allowed territorially integrated social structures and processes to be re-ordered and subordinated to functional patterns of organization.

I have already cited several components of Bookchin's argument; here it is worth quoting Merrington's basic theme length :

> Indeed the autonomous development of commercial capital, which is based on price differentials between separated markets and spheres of

production (buying cheap and selling dear) is «inversely proportional to the non-subjection of production to capital». Its externality, vis-a-vis production, is the very condition of its existence, since it interposes itself as «middleman» «between extremes which it does not control and between premises which it does not create». Merchant capital can only redistribute surplus value by windfall profits : hence its key role in the original accumulation of capital. But it cannot be a source of a permanent, self-reproducing accumulation. While it has a key preparatory role, together with its «domestic» forms of usury, speculation on scarcity, etc., it cannot play a determinant, endogenous role in the transition (to capitalism).

These considerations enable us to define more precisely the unity/opposition of towns and urban «capitalism» in the feudal mode. The «capital» and «markets» on which feudal growth was based were in no sense the linear ancestors of the capitalist world market. It is wrong to interpret the «freedom» of the medieval towns in a one-sided, unilateral sense outside the feudal context which both determined the 'externality' of this freedom of merchant capital and defined its limits. The towns' autonomy was not that of a «non-feudal island» (Postan); its freedom and development as a corporate enclave was not «according to its own propensities» as in Weber's historicist formulation. It was grounded on and limited by the overall parcellization of sovereignty, based on the coincidence of political and economic relations of subordination/appropriation which defined the feudal mode. It was the existence of this corporate urban autonomy as a «collective seigneur» within a cellular structure based on sovereignty «in several degrees» that precisely encouraged the fullest development of merchant capital in the medieval town. Hence urban «capitalism» was both internal and external to the feudal mode — or, more precisely, the former was the condition for the latter. The «internal» versus «external» terminology of the Dobb-Sweezy debate should be reinterpreted in this light. The «opposition» of these towns was an opposition of economic-corporative spheres of sovereignty : this must be seen as an element as internal to feudalism as the rise and decline of the seignorial economy — indeed as defined by this coexistence. Far from being immobile, let alone exclusively «rural», feudalism was the first mode of production in history to allow, by its very absence of sovereignty, an autonomous structural place to urban production and merchant capital (Merrington, 1975: 77-8, emphasis in the original).

It might be argued here, then, that while the medieval commune represented a non-antagonistic historical contradiction, in Friedmann's terminology (refer back to note 1), the rise of the bourgeois city amounted to its transformation into an antagonistic contradiction. This transformation was achieved through (1) the generalization of commodity-production relations, which helped destroy the economic basis of seignorial production; and (2) the competition among towns to increase their limited monopoly markets. Such increases could only occur, before the rise of entrepreneurial competition, by geographic expansion of sovereignty, which partially explains the urban bourgeoisie's willingness to cooperate with royal pretenders against the fragmented feudality. As well as guaranteeing corporate privileges, through royal protection, creation of a broader national market allowed horizontal expansion of functional economic ties. As for the aspirants to monarchy, alliance with the urban middle classes provided their only possible domestic base of economic power, vis-a-vis the territorial claims of the landed aristocracy. But according to both Bookchin and Merrington, the political ascendancy of the nation-state, as well as the development of true entrepreneurial competition and genuinely national markets, could not be based on «second-handprofiteering». The limits of corporate monopoly and the bonds of fragmented (feudal) territorial sovereignty could only be genuinely broken by a universal triumph of commodity relations (i.e. the market). And ironically, this came about through the spread of capitalistic production relations in the countryside — the only place where abstract capital and labor were free to compete against the ascriptive political economy of the guilds.

4 Industrial cities, the nation-state, and empire

Emergence of the nation-state and its establishment as the principal global expression of political sovereignty was an immensely complicated process. It took place over a period of some four hundred years, and involved a confusing quilt-work of alternating alliances and struggle between the urban middle class monarchy, several segments of capital (i.e. landed, merchant, industrial and finance), the peasantry, and the urban proletariat.

The results of these wars, civil wars and revolutions, which reached a hiatus after 1870, were hegemony of the bourgeois State, formation of the proletariat, the ascendance of finance capital, and complete dominance of the city over the countryside. Capitalist economic relations and State power all but destroyed the territorial foundations of feudal society and the mercantile city; although their remnants linger on, periodically asserting themselves at the first signs of weakness.[10] Territorial society was superseded by a new mode of functional relations between people, their work and their environment, as well as a new set of «urban» production relations between town and countryside.

> With the advent of machine production (the labor process itself) is qualitatively altered; capital seizes hold of the real substance of the labor process, dynamically reshaping and diversifying all branches of production by the technical-organizational transformation of the productive process. The removal of all fetters on the mobility of labor and the separation of one secondary process after another from agriculture (given the corresponding revolutions in transport) opens

the way to an accelerated, permanent urbanization based on the `concentration of the motive power of society in big cities' (Marx) and the subordination of agriculture as merely one branch of industry. The dominance of the town is no longer externally imposed : It is now reproduced as part of the accumulation process, transforming and spatially reallocating rural production «from within». The territorial division of labor is redefined, enormously accentuating regional inequalities : far from overcoming rural backwardness (seen as a legacy of the past, as in Smith), capitalist urbanization reproduces it within specific regions, on a more intensive basis. The creation of the «reserve army» of cheap labor and the rural exodus could scarcely be seen as «progress» from the rural standpoint (Merrington, 1975: 88).

This emerging pattern, however, contained the seeds of its own destruction : the (verticle) relations between capital and labor, and the (horizontal) relations between urban and rural. These two «moments» of capitalist industrialism introduced new historical forms of the contradiction between territory and function, and after less than half a century (1871-1914), were themselves set reeling in violent transformation.

By the second half of Period 4 (the «Mercantilist City»), as already noted, the ascendant middle classes began to form self-interested alliances with representatives of the nascent absolutionist State. While the violence of open revolt actually festered among the rural peasantry, this was a secondary sign of structural crisis within the old order; later all the traditional estates would find themselves aligned against the first modern revolutionary class, the bourgeoisie, and the proletariat which followed in its wake.

Mercantile cities had already begun to spur the break-up of territorial society : «the expansion of the market from a local or regional to an international scale occurred at a tempo that gravely disrupted the harmony of the commune» (Bookchin, 1974: 53). Trade and the renewed formation of functional linkages between various cities, began a process of sectoral specialization from place to place (based on historical «industrial mix», strategic location on the trading network, and local resources). Perhaps more importantly, interregional and international trade led to specialization of labor according to different stages of the production and distribution process, i.e. specialization within and between guilds, and the separation of manufacturing and merchandizing activities. The implications of such a division of labor were immense. As early as the 13th century there were signs of it among Flemish towns, where imported wool from England provided the raw materials for an increased supply of textile goods. Increased consumer demand (brought about by rapid population growth) as well as technological innovations — which made it possible to use unskilled labor to go into real entrepreneurial competition with weavers' guilds — led to the «putting-out system» and exploitation of peasant wage labor in the countryside. Footloose merchants would provide wool, collect the peasants' cloth for a pittance, and re-sell it dearly elsewhere. This helped monetize the rural economy, weakening the material ties of feudal relations by ending the primacy of the guilds and urban-based merchant capital. It also provided the necessary conditions for emergence of the separate workplace or factory, but full development of the «factory system» had to wait for more fertile ground, in England.

The alliance of town merchants and royal personages was a formidable combination; with steady deterioration of the seignorial economy, the royal cause gained increasing momentum. The great riches of empire provided an overwhelming boost in prestige and monetary wealth to the royal-bourgeois alliance, first bringing precious metals to kings and merchants to pay armies and hire runaway serfs — and then starting the «free» flow of other raw materials for processing and manufacture. By the beginning of the 16th century, France, England and Spain were United Kingdoms, and the «age of absolutism» followed quickly on their heels : Henry VIII (1509-1547), Elizabeth I (1558-1603), Louis XIV (1643-1715), *le Roi-Soleil*. This juxtaposition of royal centralism, empire building, merchant wealth, and the beginnings of industrialism was no accident : they were, of quintessentially linked. The landed aristocracy was made subservient to royal power and merchants rose to the prestige of a new peerage, created by fiat.[11]

The monarchy soon found itself at odds with a new alliance of capitalists, however; the remains of landed wealth — which was partially renewed by merchant investments in «property» (Merrington, 1975) — and the great commercial and finance brokers, the «real» power behind the throne. These two groups, aided and abetted by early industrialists, were the makers of the American and French Revolutions, and the creators of parliamentary government. Their alliance also outlived its usefulness, though, as «neo-feudal» country squires found their continuing source of influence in an increasingly industrial order tied to the fortunes of monarchy. Cromwell's Parliaments (1649-1660), the insurgent plantation owners of 1776, and the revolutionary Second Estate of 1789 were a far different lot than the reactionary (royalist) provincials of 1830 and 1848, or the rural pillars of the American Confederacy in 1860. (The US Civil War, while open to other, more romantic interpretations, can only be rationally understood as a confrontation between several different segments of regionally localized American capital, within the broader context of British imperialism.) These changes were tied up with the expansion of industry and its subsequent «capture» by finance capital, the roots of which take us back to 18th century England. The enclosure movement and putting-out system of manufacture grew up together in the English countryside. They were the foundations of the textile industry, which, later, switching from wool to colonial, slave-produced cotton, marked the real beginnings of the «industrial revolution» (urban-industrialization) in Lancashire. The State set the stage for these transformations through politico-legal action :

> In Britain no legislation to expropriate large property was necessary or politically feasible, for the large landowners or their farmers were already attuned to a bourgeois society. Their resistance to the final triumph of bourgeois relations in the countryside — between 1795 and 1846 — was bitter. However, though it contained, in an inarticulate form, a sort of traditionalist protest against the destructive

sweep of the pure individualist profit principle, the cause of their most obvious discontents was much simpler : the desire to maintain the high prices and high rents of the revolutionary and Napoleonic wars in a period of post-war depression. Theirs was an agrarian pressure group rather than a feudal reaction. The main cutting edge of the law was therefore turned against the relics of the peasantry, the cottagers and laborers. Some 5,000 «enclosures» under private and general Enclosure Acts broke up some six million acres of common fields and common lands from 1760 onwards, transformed them into private holdings, and numerous less formal arrangements supplemented them. The *Poor Law* of 1834 was designed to make life so intolerable for the rural paupers as to force them to migrate to any jobs that offered. And indeed they soon began to do so. In the 1840s several counties were already on the verge of an *absolute* loss of population, and from 1850 land-flight became general (Hobsbawm, 1962: 185).

A considerable volume of social overhead capital — the expensive general equipment necessary for the entire economy to move smoothly ahead — was already being created, notably in shipping, port facilities, and the improvement of roads and waterways. Politics were already geared to profit. The businessman's specific demands might encounter resistance from other vested interests, and... the agrarians were to erect one last barrier to hold up the advance of the industrialists between 1795 and 1846. On the whole, however, it was accepted that money not only talked, but governed. All the industrialist had to get to be accepted among the governors of society was money (Hobsbawm, 1962: 49).

The enclosure movement and its accompanying devastation of rural life was one of the most brutal rational acts ever perpetrated by human beings upon their fellows. Indeed, as Bookchin points out, its earlier appearance had been a major inspiration for Thomas Moore's *Utopia* :

But this (i.e. idleness of the armies maintained by the absolutist State) is not the only reason which drives men to steal. There is another one more peculiar to you English in my opinion. 'What is that ?' said the Cardinal. 'Your sheep,' I said, 'which are normally so gentle and need so little food. Now (so they say) they have begun to be so ravenous and wild that they even eat up men' (Moore, 1516: 1 3-14).

Whole families were set to wage labor in their miserable cottages. The handloom weavers became the first, rural protoproletariat; until they too were destroyed by the new urban factory system, built on the surplus value extracted from their own labor. Their futile resistance — the Chartist revolt of summer 1842 — only got them shot, as the soldiers of the «King» were sent to protect the steam boilers, power looms and factory buildings of the industrialists. The age of «free» industrial capital was born. It subdued the countryside, and after it had ruined the communal guilds of the towns, it moved in to create the industrial city. Thus, everyone was forced off the land, leaving it «free» as well for the advent of genuine factory-system agriculture. And in the cities, human labor had become a salable commodity (almost a «free» good), capital was king — steam power its prostituted queen :

> *Devant de tels temoins, O secte progressive, Vantez-nous le pouvoir de la locomotive, Vantez-nous le vapeur et les chemins de fer.*[12]

Working and living conditions in the early industrial cities provided the first tinder for revolutionary ideas in the 19th century : utopian socialism, anarchism, communism. They were also the seed beds of reformist social action, such as urban and regional planning[13]. Important as these movements were historically, however, they are somewhat tangential to the current discussion.

Town and country relations were significantly impacted, though, by other, superficially exogenous developments : (1) the growth of empire, and (2) the resultant dominance of finance capital. The industrial revolution was, among other things, an immediate, palpable result of the infamous «triangular» Atlantic slave trade. Blacks from West Africa were ferried across the Atlantic, where they were sold into perpetual bondage to work the Caribbean, Hispanic and North American plantations. Cotton — planted, harvested, ginned and baled by slave labor — was then loaded on eastbound ships (also by slave labor) and sent to Liverpool. In Lancashire the cotton was used in abominable sweat shops — i.e. factories — to make textiles, which were, in turn, exported back to Africa and America — as well as other European countries and, importantly, India. (Laborers who would not cooperate were given «transportation» in the Royal Merchant Marine, where they became genuine unpaid slaves themselves, to provide motive power when the winds failed and free muscle power in Australia and New Zealand.) India, the jewel of the British Empire, was penetrated in Bengal by Plassey of the East India Company (1757), and then consolidated militarily and politically by Clive and Hastings (1760-80). But she was brought to her knees economically by the forced importation of Lanashire textiles, which ruined a traditional industry that had earlier supplied much of pre-industrial Europe's demand for cloth.[14]

Although the cotton trade was absolutely essential to the rise of industrial capitalism, it was also responsible for growth of empire and the eventual dominance of finance capital and the «imperial» capital cities over industry. As Hymer had observed, after the 1870s international capital had become a unified force, centered and strategically directed from London[15]. A double role reversal seems to have taken place within the dialectic relations of industrialization and empire building. The role of the national capitals — i.e. the royal cities — had been that of a merchant prince at the beginning of Period 5 : they had created the national markets which provided the functional integration underlying kingly rule. But as Braudel has argued :

> The obvious fact was that the capital cities would be present at the... industrial revolution in the role of spectators. Not London, but Manchester, Birmingham, Leeds, Glasgow and innumerable small proletarian towns launched the new era. It was not even the capital accumulated by eighteenth-century patricians that was to be invested

in the new venture. London only turned the movement to its own advantage, by way of money, around 1830. Paris was temporarily touched by the new industry and then released as soon as the real foundations were laid, to the benefit of coal from the north, waterfalls on the Alsace waterways and iron from Lorraine. Sebastien Mercier's Paris was also the end of a material world. That which was born of the middle-class nineteenth century was worse perhaps for the working classes but it was no longer to have the same meaning (Braudel, 1973: 440).

This world of industrial cities at the turn of the 19th century was based on extension of the factory principle of organization to the entire fabric of urban social relations and physical development, and then to city/country relationships (Bookchin, 1974: 51-6). Colonial development during the first half of the 1800s, however, consolidated unprecedented wealth and political power in the hands of London financiers. Empire acted as an enlarged «national periphery», with resource flows directed toward England and industrial goods sent out to the colonial areas. Despite the importance of manufacturing there was little or no «direct overseas investment» as we know it today; overall market coordination was done through imperial commercial interests, closely tied to the very center of government. In effect, the flow of people, capital, resources and industrial goods was directed from the seat of empire. Even after Great Britain lost its imperial monopoly, in the last three decades of the century (Hymer and Resnick, 1971), the basic structure of the system continued, with other national variants centered on Paris, Berlin, and finally, Tokyo. In a very real sense, the self-assured (if introspective) flowering of European high culture and science during the latter Victorian period and Edwardian era — le Fin du Siecle — was based on the temporary re-emergence of an expanded territorial society. Functional forces had unified the national state and pushed it to the very ends of the earth — but it still held in tact. Financial interests ran the imperial government — indeed, became the government — and everything turned in fixed orbits around the imperial city. National monopolies based on banking and finance were the true royalty, not the King, Kaiser or Emperor (Lenin, 1917). Peoples and activities which lost their own momentum were drawn to the center : «... London, that great cesspool into which all the loungers and idlers of the Empire are irresistibly drained» (Conan Doyle, 1936: 6). London and Paris — the nerve centers of the two most far-flung national empires — grew to exceedingly large dimensions by the 1870s. But with the entry of impatient new national competitors, the imperial cities and their expanded territorial hinterlands were locked on a collision course. Territorial empire was impervious. There were lines drawn around what belonged to whom. Each group of nationally-based financial interests had their label attached to their property. If late comers wanted a share, and there was little of the world left to divy out, then someone else had to loose part of their allotted portion.

And this was, of course, the reality behind the fatal explosion of 1914. Germany's unification in 1871 had established a new, aggressive imperialism in the heart of Europe. Bismarck's imperial designs were founded upon an ultra-

modern capitalist industrial base — -more than competitive with anything in
Britain or France — and the pent up national-cultural feelings of the German-
speaking peoples of north-central Europe, who had been the pawns of
European politics for three centuries. The authoritarian Prussian State created
the first truly contemporary nation, based explicitly on an alliance between
Kaiser, big capital and the Junker military class. In retrospect, it may even be
questioned whether Kaiser and Reichstag were not largely transparent symbols
of conformance to earlier European traditions.

The newly united Germany of 1871 displayed a particularly balanced
economic geography. Spared the centralism of absolutist rule, no royal
encampments had ever grown up to drain the people, skills, capital and material
wealth of the country into one parasitic entrepot — as had occurred in France
and England (Braudel, 1973: 436-40). Capitalism, beyond the level of internal
mercantilism, of the sort typical towards the end of Period 4 in the more
progressive countries, had never really established itself in Germany, without the
political umbrella of monarchy. Various regional cities grew up under the
protection and patronage of a resilient feudal aristocracy, exercising exploitative
but local control over the surrounding countryside.[16] Territorial structures of an
earlier mode continued to bound and shape capitalist development.[17] Overseas
expansion into such areas as Tanganyika and Southwest Africa brought
relatively little booty to commercial interests which might have strengthened
their hold over national affairs. And expansion of Dual Alliance interests into the
Balkans resulted, first, in entanglement with the Crimean wars of the '80s and
'90s, and, then, involvement in all the muddled plotting of the Triple (Dual)
Alliance and Triple Entente. The decaying Habsburg monarchy — Germany's
partner in the Dual Alliance — and its fears of Serbia finally brought the world
of the imperial city to the brink of disaster in June 1914.

At the outbreak of World War I there were two predominant types of large
cities in the industrialized world, the imperial capitals such as London and Paris
and the sordid industrial centers like Birmingham, Manchester, Lille, Frankfurt
and Chicago.[18] Industrial capital and its physical manifestations — the industrial
cities — were subordinated to the logic and strategies of the nation-state,
dictated and organized from the swollen imperial capital, where clerical activities,
finance, and the other bureaucratic affairs of empire took their place along side
more general urban services and industry. The countryside, on a nearly global
basis, was under the rein of the metropolis, as factory relations came to dominate
rural as well as urban living. In fact, rural life, especially in the United Kingdom,
had become an endangered species, at the time of the First World War over 75
percent of the UK's population was classified as urban and metropolitan London
counted 6.5 million inhabitants (Hall, 1966: 18; 1975: 34). According to
Merrington :

> The tendency of capitalist enlarged reproduction is to revolutionize
> all fixed divisions of labor (in contrast to *manufacture*); it recomposes
> the labor force by constant `variation of labor, fluidity of function and
> universal mobility,' undermining the existing relation between the
> worker and his job — the use-value of his work — tending towards

the subordination of universal indifferentiated labor at the service of accumulated dead labor (constant capital), bringing the countryside into the factory and the factory to the countryside in its restless search for fresh manpower. In this levelling and mobility of the labor market, the «factory city» already prefigures the sprawling conurbation, the «megalopolis» of the 20th century, the absolute negation of the «city» to humanist critics and planners. The capacity of fully socialized capital to appropriate earlier utopias based on the ideal of a balanced environment, to transform them into a «technical matter at the service of neo-conservative established powers», is demonstrated by the «garden city» ideal (derided by the Fabians as a pipe-dream in 1898) and its reality in the *planned deurbanization of the metropolis,* dissolving the city into the «urban region» in the town and country planning of the 20th century. («Town and country», wrote Howard, «must be married and out of this union will spring a new life, a new hope, a new civilization».) The mobility of mature social capital presupposes this capacity to reconstitute the town-country division on an ever-renewed basis; while the town-country opposition becomes that of *agricultural versus industrial prices* — an increasingly political, rather than market-price determination (subsidies, quotas, price-fixing) since the need to control the cost of reproducing labor power — the price of reformism in our own day — clashes with the interest of agricultural producers (Merrington, 1975, 88, emphasis in the original).[19]

5 Megalopolis and the multinationals

It is now generally accepted that the twenty-one year interlude between 1918 and 1939 was only a breathing spell in the hostilities which began in 1914. Forces were regrouped and technological advance spurred a new, more terrifying round of armament. Despite the unduly harsh character and longevity of the occupation of the Rhineland, and the chauvinistic jingoism and brutality of fascism (especially Nazism, under Hitler's depraved leadership), World War II was, in fact, a continuation of the earlier imperialist conflict.

The last three-quarters of a century have witnessed several crucial developments. First, in the two decades which separated the World Wars, the imperial empires of England, France, Belgium and Holland began to crumble from within. Social forces in the dependent colonial areas had developed to such a point that domination from the center, from the imperial city, was increasingly unrealistic. The formation of an embryonic bourgeoisie, in both white settler colonies such as British East Africa and French Algeria, and «native» areas, like India, Nigeria and Indonesia, marked a critical turning point. And the emergence of a peripheral middle class was, necessarily, accompanied by the fledgling beginnings of a local proletariat or working class. Such developments undermined the material foundations of empire, as represented by the classical imperial division of labor under the hegemony of finance capital. The precursors of Period 6 were already in formation, for productive forces could not be artificially restrained indefinitely by an imposed strategy of «capital deepening». It can be argued that World War II merely speeded up the inevitable transition.

An important related trend was re-emergence of industrial capital from the control of financial interests. The formation of mammoth joint stock companies, verticle and horizontal monopolies, trusts, cartels and holding companies of various sorts was a «natural» outgrowth of the dynamics of the capitalist enterprise (Hymer, 1968, 1972; Hymer and Rowthorn, 1970). It had its most fundamental manifestation in the «young» capitalist economies of Germany and the United States, where finance and central government power had never been so firmly wedded as in France and Britain. These developments had been catapulting ahead since before the turn of the 20th century, and the stock market crash of 1929 only put a temporary end to the wheeling and dealing of the «Captains of Industry» (see, Veblen, 1904).

The enormous infusion of public capital into industrial production during both world wars, in Germany, the United States (and Japan), as well as the unavoidable decline of more traditional capitalist structures, mentioned just above, set the stage for a complete realignment of functional and territorial relations, which was to mushroom in the post-1945 era. The decline of territorial power (empire) and the rise of «internalized» functional economic organization (the multi-national corporation), marked the death knell for both the State/finance capital alliance which had dominated the world economy for 80 years (1870-1950), as well as town/country relations based on the historical contradiction between imperial metropolis, industrial city and (national) world periphery.

The rapid collapse of European imperialism after World War II was accomplished, for all practical purposes, by the early 1960s. Political independence in one colonial area after another broke up the outdated territorial empires, and left in their stead literally dozens of shaky, new nation-states, under the would-be governance of an aspiring local middle class. As in Carolingian Europe a thousand years earlier, attempts were made to preserve appearances of the waning territorial order, through such shadowy, atavistic constructs as the British «Commonwealth» and the French «Community», but the reality of military and economic power lie elsewhere. For the first thirty years following 1945 (outside the vast Asian mainland comprised of the former Soviet Union and the People's Republic of China), American military power imposed a new *Pax Americana.*

In this world of imperial retrenchment and aspiring nationhood, a new economic imperialism was born, based on direct foreign investment by American industrial corporations. Supposedly inviolate national sovereignty and free international trade were the cornerstones of the new order (although reluctance in accepting the latter principle quickly brought into question the former as well). First through the Marshall Plan rebuilding of Europe and then on an even broader international scale, the multinational corporation, with functional economic activities divided among many dispersed production units (without regard for national boundaries), established the first truly global division of labor. As the shattered industrial economies of Germany, Japan, the UK and France reasserted themselves, they too fielded competing corporate investors — as was only «fair» under the prevailing rules of the game. But the initial nationality of corporate enterprises has shown itself to be a matter of trifling importance, for by the 1990s functional organization of economic activities and the territorial interests of nation-states have proven highly contradictory. This is demonstrably true for the United States itself : witness the low rate of capital investment and profits brought «home» by American multinational corporations, or the explicit targeting of the American economy for punishment by supposedly «American» multinational oil companies during the two «Oil Shocks».

In point of fact, functional economic power is becoming substantially divorced from territorial society, just as clan society was disenfranchised by the ascendance of territoriality. The «multinational» corporation has been transformed into a «transnational» enterprise, owing allegiance to no territorial state and effectively controlled by none. This represents a new form of economic feudalism.

The implications of escalating transnationalism for national societies are just beginning to emerge, and they are indeed profound. Highly polarized social and geographic patterns of development are unavoidable under such a system. Both capital and relevant technological information are internalized *within* the transnational corporation. Territorial units of society and government below at least a continental — if not global — scale are helpless, especially at the important level of subnational jurisdictions : without effective taxing authority, monetary controls, migration and investment laws, fiscal independence or the threat of military force. They are even more prostrate than during the period of national market consolidation, because functionally organized «trancos» have none of the worries of political sovereignty and, thus, no politically dictated welfare functions. If political, economic, social or environmental conditions in a particular area become overly worrisome or costly, the transnational corporation just packs up and moves on.

Relations between town and countryside are, of course, undergoing yet another transformation as a result of the new functional dispensation. The industrial city, as well as its larger imperial counterpart, had undergone spectacular growth during the final decades of Period 5. What Patrick Geddes had called «conurbations» and Lewis Mumford «dinosaur cities» had grown up around the initial nucleus of the factory town and the «royal camp». The migration of almost whole national populations into a few dynamic urban centers was the outstanding demographic and economic phenomenon of the century following consolidation of bourgeois power in and around 1848. Economic subordination of both city and countryside to «urban» factory relations left few employment opportunities in the rural areas, and people by the millions came pouring into the metropolitan centers.

After the advent of steam-powered public transportation, cities were loosed from the physical growth restraints of natural human muscle power; only the rich and privileged had been able to afford the luxury of horse-drawn carriages. The wild-fire expansion of cities like London and Chicago after the introduction of urban commuter railways has been a subject of much investigation and acclaim by students of metropolitan growth. And this initial serpentine outpouring of urban-like development was literally dwarfed by later expansion, after the generalized spread of individual automobile ownership following the Second World War. The urban factory spilled out to annihilate tens of thousands of square miles of formerly agricultural land, first creating the vast metropolitan agglomerations described and modeled by the Chicago School of Urban Ecology, and then going on to call forth Gottmann's (1961) megalopolis : entire «metropolitan» corridors sprawling with segregated, low-density, urban-type land uses : housing, factories, retailing centers, warehouse districts, government buildings, schools and hospitals; reaching its zenith in the non-stop freeways and strip-commercial development of Southern California — an island on the land (McWilliams, 1973; Soja and Scott, 1986). Rondstadt Holland and the Tokyo - Osaka corridor are clones.

It would be the most fundamental error, however, to equate such development with urbanization or final destruction of the «rural» by the «urban». This is the «localized» physical manifestation of functional economic relations triumphant — the mega-factory :

> the factory is the locus of mobilized abstract labor, of labor power as a commodity, placed in the service of commerce as well as production. Accordingly, the term applies as much to an office building and a supermarket as to a mill and a plant. Once the factory becomes an element of urban life, it takes over the city almost completely (Bookchin, 1974: 51).

In this factory without walls or limits, little residual trace can be found of territorial society, or, for that matter, the city. There is no «communal densification» as was the case from western antiquity onwards in cities; some relative densification, yes, but no community. Outside of marginal immigrant enclaves — like the Mexicans or Koreans in Southern California, the Puerto Ricans in New York, the Algerians in Paris and Marseille, the Indians, Pakistanis and Caribbean Islanders in South East England — only the deteriorating nuclear family provides any semblance of human social relations outside the marketplace. And the «monogamous» family, too, is in a stage of irreconcilable decay; as male dominance and female exploitation give way to a new sort of economic «freedom». Everyone and everything becomes subject to money exchange. Everyone works in the factory. (Even food preparation is becoming rapidly transferred to the corporate sector.) And everyone increasingly has the right to experience complete alienation of their labor into strict commodity relations. As men and women genuinely achieve the social status of independent, «equal» monads, pairing couples no longer bother to assume a common name; and there is nothing remaining of the earlier gentile ties of the extended family to offer identification or identity. Economic and technological progress have made the nuclear living unit unnecessary, and only the increasing concentration of property wealth in corporate hands — even to the point of individual housing units, etc. — prevents the readoption of some new form of «mother-right», to identify parentage and pass on accumulated wealth.

Sheer territoriality has also shrunken to near insignificance. «Neighborhood» units have been reduced to nothing more than street addresses and postal codes. Residential mobility is almost universal; most people never have time to learn even the «neighbors»' names. And as Melvin Webber (1963; 1964) pointed out already 30 years ago, interpersonal relations are increasingly oriented toward professional groupings and other job-related (i.e., functional) ties. We are more apt to have dinner with someone from across town, from another city, or, even, another country than our downstairs or next door neighbor. The neighborhood, the city and the metropolitan region have all become strictly throwaway containers.

Government — the State — has also abandoned the city. With unification of the nation-state, government had already moved from the level of the city-region (or its imperial extension) to another, less easily identifiable geographic scale. Economic matters and other life and death decisions for communities and individual members took place in London, Washington or Paris. And in the case of an increasing number of countries, the historical American model of an «artificial», new town capital — with few if any traditional rights of self-governance or customary citizenship — is being established, exclusively to house the transient bureaucratic population and insulate the State from unseemly or troublesome neighbors. And while the State continues to initiate territorial economic policies, such as urban-regional planning and collective urban consumption, these only impact local populations if and as they meet the needs of functional economic interests.

Finally, on the economic side, even the industrial factory is bailing out of the «city», now that it has subjugated everything to its own logic through transnational privatization. First industrial production moved to the periphery of the metropolis, to escape the congestion and other negative externalities of its own making. Then it began a forced march back into the countryside from which it had arisen, in search lower wages, land values, taxes, and other reduced social obligations : i.e. the bonds of territorial community. Needless to say, the empty, mechanized countryside demands even less social responsibility than the de-urbanized metropolis, and enfeebled local jurisdictions in rural areas allow themselves to be bought very cheaply.

In sum, the «urban fields» of the late twentieth century (Friedmann and Miller, 1965) are being set adrift economically. First industry moved to the outskirts, then on to the countryside. Now transnational capital has almost completed the move of secondary industry to the far edges of the world periphery, where labor is all but free and government is frequently too unstable to provide more than token resistance to the most exploitative, neo-sweat shop arrangements. (Typical nationalization and majority ownership procedures are little more than large ante bribes.) For the first time in decades, population trends in the developed world have reversed themselves. In the United States, for instance, small cities, towns and open country communities located outside large «cities» and their suburbs have, since 1970, recorded the nation's fastest growth rates (Fuguitt et al., 1979). With deindustrialization, large cities are well on their way to affording two and only two kinds of employment in the all-encompassing corporate sector : clerical-R&D-management and «urban» services (for both households and industry) (Bluestone and Harrison, 1982). Only a relatively small proportion of the population is actually involved in the former category of activities (not including direct government employees), so much of the exchange value created in megalopolis is directly attributable to such «non-productive», self-fed activities as personal services (and, of course, speculation in real property).

Functional economic activities are straining today to divorce themselves even more completely from the social matrix of territorial community, as well as beginning their final assault on the ties of sex and the family. The contradiction between territory and function, however, rather than being shattered by total rupture of the territorial community (at the limit, an ontological impossibility), is set into even bolder relief. Areas which have been the hardest hit — nation-states, subnational regions, and large, incorporated «cities» — are already engaged in frequently violent confrontation. Others are slowly queuing up to join their ranks, as their own supposedly favored relations with transnational capital are threatened for one reason or another (see, Friedmann, 1986). Struggles within the Group of Seven over the Uruguay Round of Gatt have their origins here.

What, then, about the antagonistic contradiction between city and countryside ? In the more «developed» regions of North America, Western Europe and Japan (see Douglass, 1987), I would submit that it has virtually disappeared. The factory — meaning, functional economic relations — has come close to burying both concepts in the slag heap of historical definitions. Their palpable material and ideological reality has been stretched to the substance of a wraith. Both have lost almost all of their «territorial» attributes, properly speaking. One interpenetrates the other with impunity in both social and geographic terms, and both have been robbed of all but the shabbiest remnants of their most characteristic historical features. Industry is «located» in neither, as it were; they have both become its containers. There are industrial factories in the countryside and agricultural factories in the «city». Town and countryside, persist as only hollow spatial forms, merely historical «residuals». Currently there exist only the most immature territorial institutions at the international level, however, and functional economic power remains elusive moving from one boardroom to another, from one transcontinental flight to the next. As yet it has no new global army or global capital despite recent developments within the United Nations. At worst, in the traditional industrialized countries of the west, town and countryside lie in an almost dormant non-antagonistic opposition; both emaciated by functional economic relations, for which each, in its turn, has served as the leading vehicle.

If this is hyperbole, I would argue that it is only a matter of timing and clear-cut dominance. Transnational power is in its adolescence; many of the tendencies described in the last few pages are in disparate stages of development. And all are locked in unavoidable struggle with historical antecedents and alternative possibilities — dominance is a matter of human will and social organization, not just political, economic and technological potentials.

Recent transformations of the contradiction between town and countryside have created a much more probablistic situation then has existed for several centuries, as the now traditional alliance between cities, capital, industry and the State has flown asunder. The attempted creation of multinational regional blocs, such as the European Community and the North American Free Trade Area, may enforce a new continental level of territorial closure on functional economic integration. This is yet to be seen, however.

Notes :

1 Most people educated in the western tradition associate dialectical thinking with Hegel and Marx; some are able to carry the dialectic method back to its occidental origins in Socratic dialogue (see for instance, Weaver, 1985). Friedmann's conceptualization of the contradictions between the urban and the rural draws on this intellectual heritage, but underlays it with a radically different ontological scheme borrowed from Chinese Taoist/Maoist thought. For Hegel and Marx contradictions are a structural juxtaposition of mutually antagonistic social tendencies and forces, working their way out in history through a linear progression of thesis, antithesis, and synthesis. In Hegel's idealist formulation, drawing on Kant and the Greeks, this is the «Idea» weaving «History» and the future. In Marx's supposedly materialistic epistemology, contradictory social forces are locked in real struggle in the world of things, people and events. The East Asian tradition conceives of contradictions as the relationship between Platonic-like pure forms, intermingled in a cosmological unity of opposites, such as the male and the female : the yin and the yang. This relationship of unity and conflict, dominance and dependency, can take many concrete historical forms in the world, and it does not necessarily lead to any particular linear trajectory in mundane affairs. Existence is cyclical not progressive. Friedmann identifies the relationship between city and countryside as a cosmic contradiction, a fundamental unity of opposites. In their essential ontological nature, he argues, the urban and the rural are defined in terms of one another and their interrelationship of unity and conflict, dominance and dependency. In actual historical settings this contradiction can take different concrete forms. Sometimes they exist in a state of open conflict — antagonistic historical contradictions — which sets off an Hegelian linear progression of thesis, antithesis, and synthesis, eventually achieving a new and higher unity. At other times they are intermingled in a relatively benign non-antagonistic historical contradiction, a unity of opposites existing in a kind of temporary historical equilibrium. In terms of the argument presented in this chapter below, city and countryside might be thought to coexist in a non-antagonistic historical contradiction at the beginning of Period 2 and the end of Period 3 in Table 1 (the Classical City of Mediterranean Civilization and the Towns and Burgs of the Middle Ages, respectively). Periods 1 and 5, the Temple/Palace of Occidental Antiquity and the Imperial City of Industrial Capitalism, represent eras of antagonistic historical contradictions. My disagreement with Friedmann's position, and the thesis of this paper, can be simply stated. I accept Friedmann's «Pacific mind set», mixing eastern and western ideas. I believe his formulation provides an important step forward in analyzing the relations between the urban and the rural. Substantively, however, Friedmann argues that the urban has always dominated the rural historically, either in an active (antagonistic) or passive (non-antagonistic) manner. I believe this is mistaken. Such a position refutes the logic of his basic Taoist construct and overlays it with an unrealistic historical progression of increasing urban domination. This seems to me overly western and Hegelian. I will attempt to demonstrate in what follows that the history of rural/urban relations appears to be much more cyclical, and that both the countryside and the city have taken turns in establishing the fundamental dominance/dependency terms of their contradictory interrelationship.

2 Table 1 represents the synthesis of a great deal of information. I have made no comprehensive attempt to document the lineage of the various ideas incorporated. This was fundamentally a matter of expediency, because to have done otherwise would have represented a monumental undertaking in and of itself. My principal sources in construction of Table 1 are cited in the text. I also made substantial use of

many standard references on urban geography and the history of urban design. The general interpretation of urban/rural relations for Periods 1-4 on Table 1 was significantly influenced by Bookchin (1974, 5-56).

3   Two of the most obvious possible additions to the periodization used here are : (1) the Mediterranean colonizing and empire building which characterized the latter part of both Periods 1 and 2, e.g., the Phoenician colonies in North Africa, and the spread and transformations of the Roman city with passage from the Republic to the Empire; and (2) the intermediate stages of the urban transition during the High Medieval Period in central and western Europe. In this instance, however, I think that their omission tends to clarify and simplify my central argument, without distorting its historical accuracy. Frequently, also, other periods, which overlap the major transition zones are chosen (e.g. Braudel, 1973), in order to emphasize the transformations themselves. Here, this is accomplished in the text.

4   This interpretation follows closely the arguments made by Childe (1936) and Bookchin (1974), but differs substantially from Marx (1857a, 1857b). While mentioning the experience of countries such as Mexico, Turkey and Peru, however, Marx's arguments rested primarily on the examples of the Roman Republic and medieval Europe, both of which I will touch on briefly below.

5   All dates for the remainder of the chapter are AD, unless otherwise stated.

6   I have purposely left out Claudius (41-54) from the list of maniacs and degenerate butchers, although some accounts, e.g. Tacitus (1956), suggest that this could be an error, at least as concerns other members of his immediate household.

7   Urban agriculture, however, was nothing new to the medieval town. As far back as occidental antiquity, even in the most densely populated cities, substantial amounts of land — frequently within the city walls themselves — were set aside for gardening and food production.

8   The only true cities in Europe and the western Mediterranean at this time were the   centers of Islamic civilization which flowered in North Africa and Spain. Because of their complete failure to influence later urban development in Europe, however, I will not deal with them here.

9   Merrington (1974, 76-7) quoted Marx (1857b) to the effect that the specificity of feudal towns in the West was that earlier cities were really «`royal camps'», i.e. «`works of artifice created above the economic construction proper'». While on the contrary, «`the Middle Ages (Germanic period) begins with the land as the seat of history, whose further development then moves forward in the opposition between town and countryside; the modern age is the urbanization of the country, not ruralization of the city as in antiquity'». As has already been noted, I take exception to this latter assertion, insofar as it means that the countryside is really being converted into a city, but it properly underlines the striking difference in the economic basis of urbanization before and after the feudal period in Europe.

10   Both the contemporary regionalist movement in Europe and the beginnings of political regionalism and regional planning in the late 19th century owe their origins to cultural and spatial «residuals» of the feudal commune and mercantilist city. American regionalism, in New England and the Old South, was also a socio-spatial remnant of earlier production relations; this time mercantile capitalism and the transitional stage to industrial capitalism. (Interestingly, the internal colonialism of «reconstruction», attributed to finance capital by Friedmann and Weaver (1978, Chapter 2), corresponds quite neatly — 1865-1914 — with the international reign of finance capital under classical capitalist imperialism.)

11   Contemporary regionalism in Northern Ireland and Scotland can be traced historically to the changing political alliances of this period. Rise of the absolute monarchy in England and its later struggles with the ascendant middle class were inexorably linked to English union with Ireland and Scotland. Henry VIII completed the absolutist Tudor State which his daughter was to rule for almost half a century with conquest of Ireland in 1542, five years before his death. Under the Stuarts, Ulster was settled in 1610, sealing the «first» Union of the two countries, and rebellion during the reign of Charles I lead to the catastrophic Irish Massacre of 1641. The second conquest of Scotland under Cromwell in 1650 and the formalized first Union in 1652 played an integral role in establishment of bourgeois power during the interregnum (1649-1660). The Stuart Restoration under Charles II was explicitly engineered through making deals with the Scottish and Irish, which put a king back on the throne in England, in part, by temporarily undoing English hegemony over two outlying nations. The «bourgeois» monarchy which followed coronation of William and Mary (1689-1702) was, in fact, the final establishment of middle class power over «Great Britain», formed by opening the entirety of the British Isles to unified mercantile capitalist rule with Scottish (1707) and Irish (1800) Union. The concrete circumstances for union were very different, of course — Scotland taking part in the spoils of empire, and Ireland being one of its first

victims — but with exception of the loss of the American colonies (1783), this marked the beginnings of relentless expansion of the British Empire (e.g. consolidation of India by Hastings, 1780) and something over one hundred years of *Pax Britannica*. This, the fe first unified world system of political economy, was the model of classic capitalist imperialism, based on the London finance market. It provides an important component for explanation of the imperial city's unprecedented growth, as I will argue shortly.

12 Quoted from R. Picard, *Le romantisme social,* by Hobsbawm (1962) p. 202.

13 For a discussion of early metropolitan growth and its relation to the appearance of urban and regional planning, see Friedmann and Weaver (1978), Chapter 3.

14 Thus, India became underdeveloped — rather obviously, something other than a state of natural backwardness. This story — the exploitative development of classical capitalist imperialism — makes an extraordinary bit of historical reading when juxtaposed against the logic of the various substantive theories of economic development and urban/rural relations. See especially Harnetty (1972).

15 Nabudere (1977) gives a particularly interesting account of London's role in centralizing finance capital.

16 This was the world which inspired Johann Heinrich von Thunen's model of the «isolated state» (1826); probably the first explicit location model. The *Isolierte Staat* of Thunen was just that : an isolated State, and the resultant concentric patterns of economic land rents (a notion he co-invented with Ricardo) were only possible within such an environment. Hierarchy in the sense used by Christaller (1933) one hundred years latter, or Soja in his spatial universals, was only weakly developed — or else agricultural land use patterns, especially for higher value crops, would have certainly taken other patterns. While, in typically German fashion of the period, Thunen claimed his construct was a purely «ideal» or theoretical one, it must certainly have reflected some idealization of existing material relations. The fact that this basic model, perhaps because of its convenient «gravity» characteristics, has been generalized to describe all sorts of other land use and economic location arrangements is quite unfortunate and misleading. Because the isolated political unit and social formation which provided the basis for Thunen's construct have no substantial reality under industrial or transnational capitalism. Thunen's model, ordering the functional processes of economic production, was highly constrained by a lingering spatial structure, inherited from the feudal mode of production.

17 This is even true today. Because of Germany's late start in national unification and the resultant tardy formation of a national market, as well as its century-long lag in developing a modern capitalist industrial base, Berlin never reached the bloated dimensions of London or Paris. Limited overseas possessions (only a few notches above the petty American imperialism of the 19th century) and the loss of two World Wars led to Germany's partition and political and material decline of the former capital. The impact of reunification on Berlin is yet to be seen. Although the Munich-Essen corridor dominates contemporary German economic life, the Federal Republic is literally sprinkled with moderate-sized urban centers and relatively balanced regional economies. Extensive studies of the German economy by the Center for Regional Economics at the University of Aix-Marseille have demonstrated the remarkable differences in spatial structure between western Germany and France, for instance.

18 Liverpool, Chicago and New York owed much of their early growth to their port and entrepot functions within the national and international markets, but their phenomenal growth during the late 19th and early 20th century was based substantially on industrial expansion (See, Weber, 1899). New York, like Paris, was the seat of a secondary financial and banking community as well; finance capital in North America and on the European continent, however, was distinctly subservient to the moguls of the City of London.

19 Merrington makes several extremely important points in this quotation : (1) the paramount significance of broader capitalist relations, outside the sphere of secondary industry, (2) the relationship between factory production and labor, (3) penetration of both town and countryside by factory organization, (4) coalescence of the booming `factory cities' (`conurbations' according to Geddes) into megalopolis, (5) the consequent deurbanization of the city and the formation of `urban regions,' and (6) the utopianism of the 19th century `garden city' movement. Unfortunately, he also makes two rather sloppy analogies which could be the source of continuing confusion. First, both in this quotation and in the whole latter part of his essay, he differentiates between `factory' and `city', but then goes on to equate the two with little further qualification. This is particularly misleading in the present context, because the growth of conurbations and megalopolis — especially since 1945 — is only marginally related to economic activities which might reasonably be equated with secondary industry. And secondly, this lack of

clarity leads Merrington directly into the important but confusing assertion that «the town-country opposition becomes that of agricultural versus industrial prices — an increasingly political, rather than market-price determination». There can be little doubt that unequal rates of exchange between the primary and secondary sectors of the economy are commonplace, and, further, that in the USA, EC and Japan their relationship is primarily political. But the suggestion that, in the contemporary world, which Merrington is explicitly referring to, this constitutes a contradiction between city and countryside is quite mistaken. In the first place, as I will argue in the next section above, industry (in the sense of secondary activities) is currently abandoning the city (indeed, the very core countries of industrial capitalism) with all possible haste and has been for a considerable period of time. What is more, to a significant extent, both industry and agriculture (now merely a branch of industrial production, as Merrington himself argued earlier) are integral parts of the same corporate enterprises. In the market economy, there hardly exist any remaining forms of agricultural production which could honestly be called rural, and many of the horticultural products which account for an important component of the money exchange between urban consumers and the agricultural sector are actually produced within the bounds of the supposed megalopolitan city. At the very most, `agricultural versus industrial prices' today represent a conflict between different segments of capital, and, to an ever increasing degree, not even this is true; both primary and secondary industry are being rapidly absorbed into a global corporate sector, made up of oligopolistic, parallel vertical combines which control all kinds of production from first breaking of the soil to final delivery of the consumer product. Contradictions between city and countryside are irrelevant to such a system, because functional economic relations are quite divorced from traditionally recognizable territorial structures or territorial political control.

References

Bloch, M., 1961,. *Feudal Society*. Vols. 1 and 2. Translated from the French La Société Féodale by L.A. Manyon. Routledge and Kegan Paul : London.

Bluestone, B., and Harrison, B., 1982,. *The DeIndustrialization of America*. Basic Books : New York.

Bookchin, M., 1974,. *The Limits of the City*. Harper Colophon : New York.

Braudel, F., 1973,. *Capitalism and Material Life*. Translated from the French Civilisation Materielle et Capitalisme by M. Kochan. Harper Colophon Books : New York.

Braudel, F., ed. 1985,. *La Mediterranee : L'Espace et L'Histoire*. Flammarion : Paris.

Childe, V.G., 1936,. *Man Makes Himself*. Tavistock Press : London.

Christaller, W., 1933,. *Die Zentralen Orte in Sûddeutschland*. Jena.

Conan Doyle, A., 1936,. *The Complete Sherlock Holmes*, with a Preface by C. Morley. Doubleday, Doran : Garden City.

Coulanges, N.D., Fustel de 1874,. *The Ancient City : A Study on the Religion, Laws and Institutions of Greece and Rome*. Paris. Current American edition : 1980, Johns Hopkins University Press : Baltimore.

Dobb, M., 1946,. *Studies in Development of Capitalism*. Routledge and Kegan Paul : London.

Douglass, C.M., 1987,. *Transnational Capital and Urbanization in Japan*. Discussion Paper, N°. 1. Department of Urban and Regional Planning, University of Hawaii.

Engels F., 1984,. *The Origin of the Family, Private Property and the State*. Zurich. Current edition : 1977, Progress Publishers : Moscow.

Friedmann, J., 1979,. 'On the Contradictions between City and Countryside,' in *Spatial Inequalities and Regional Development* Eds. J. Osterhaven and H. Folmer,. Nijhoff : Leyden.

Friedmann, J., 1985,. *The World City*. WCP, 9. Graduate School of Architecture and Urban Planning. University of California, Los Angeles.

Friedmann, J., and J. Miller 1965,. 'The Urban Fields,' *Journal of the American Institute of Planners*, 31, pp. 312-320.

Friedmann, J., and C. Weaver 1979,. *Territory and Function : The Evolution of Regional Planning.* University of California Pres : Berkeley and Los Angeles.

Fuguitt, G.V., P.R. Voss and J.C. Doherty 1979,. *Growth and Change in Rural America.* Urban Land Institute : Washington, D.C.

Gibbon, E., 1788,. *The Decline and Fall of the Roman Empire.* Current American edition : 1963, Dell Publishing : New York.

Gottmann, J., 1961,. *Megapolis : The Urbanized Northeastern Seaboard of the United States.* M.I.T. Press : Cambridge,

Griffeth, R., and C.G. Thomas, eds. 1981,. *The City-State in Five Cultures.* ABC-Clio : Santa Barbara, CA.

Hall, P., 1966,. *The World Cities.* McGraw-Hill : New York.

Hall, P., 1975,. *Urban and Regional Planning.* John Wiley : New York.

Harnetty, P., 1972,. *Imperialism and Free Trade and India in the Mid-19th Century.* University of British Columbia Press : Vancouver, Canada.

Hobsbawm, E.J., 1962,. *The Age of Revolution, 1789-1848.* Mentor Books : New York.

Hymer, S., 1968,. 'La grande corporation multinationale,' *Revue Economique,* 19/6, pp. 949-973.

Hymer, S., 1972,. 'The Internationalization of Capital,' *Journal of Economic* Issues, 6, pp. 91-111.

Hymer, S., and R. Rowthorn 1970,. 'Multinational Corporations and International Oligopoly : A Non-American Challenge,' in *The International Corporation* Ed. C.P. Kindleberger,. M.I.T. Press : Cambridge, MA.

Hymer, S., and S. Resnick 1971,. 'International Trade and Uneven Development,' in *Trade, Balance of Payments and Growth :* Papers in International Economics in Honor of Charles P. Kindleberger Eds. J.N. Bhagwati, R.W. Jones, R.A. Mundell, nd J. Vanek,. North-Holland : Amsterdam.

Lenin, V.I., 1917,. *Imperialism, the Highest Stage of Capitalism.* Current edition : 1970, Foreign Languages Press : Peking.

Marx, K., 1847,. *The Poverty of Philosophy.* Paris. Current American edition : 1963, International Publishers : New York.

Marx, K., 1857a,. *Pre-Capitalist Economic Formations.* Current American edition : 1965, International Publishers : New York.

Marx, K., 1857b,. *Preface and Introduction to a Contribution to the Critique of Political Economy.* Current edition : 1976, Foreign Languages Press : Peking.

Marx, K., 1875,. *Critique of the Gotha Programme.* Current edition : 1972, Foreign Languages Press : Peking.

Marx, K. and F. Engels 1846,. *The German Ideology.* Current American edition : Internatio,nal Publishers : New York.

Matthiae, P., 1981,. *Elba : An Empire Rediscovered.* Translated from the Italian Elba Un Impero Ritrovato by C. Holme. Doubleday : Garden City, NY.

McWillims, C., 1979,. *Southern California : An Island on the Land.* Peregrine Smith : Santa Barbara, CA.

Merrington, J., 1975,. 'Town and Country in the Transition to Capitalism,' *New Left Review,* 93, pp. 71-92.

Moore, T., St. 1516,. *Utopia. Flanders.* Current American edition : 1965, translated from the Latin by J.A. Scott. Washington Square Press : New York.

Morgan, L.H., 1877,. *Ancient Society, or Researches in the Lines of Human Progress from Savagery Through Barbarism to Civilization.* Macmillan : London. Current American edition : 1978, New York Labor News : Palo Alto, CA.

Mumford, L., 1938,. *The Culture of Cities.* Harcourt, Brace and World : New York.

Mumford, L., 1961,. *The City in History : Its Origins, Its Transformations, and Its Prospects.* Harcourt, Brace and World : New York.

Mumy, G.E., 1978,. 'Town and Country in Adam Smith's The Wealth of Nations.' *Science and Society,* 42/4, pp. 458-477.

Nabudere, A., 1977, *The Political Economy of Imperialism.* Zed Press : London.

Palmer, T. ,1961,. *Historical Atlas of the World.* Rand McNally : Chicago.

Pettinato, G., 1981,. *The Archives of Elba : An Empire Inscribed in Clay.* Translated from the Italian Elba. Un imperio inciso nell argilla. Doubleday : Garden City, NY.

Pirenne, H., 1925,. *Medieval Cities : Their Origins and the Revival of Trade.* Translated from the French Les Villes Medievales by F.D. Halsey. Princeton University Press : Princeton, NJ.

Pirenne, H., 1958,. *A History of Europe.* Vol. 1. Translated from the French L'Histoire de L'Europe by B. Miall. Anchor Books/Doubleday : Garden City, NY.

Plutarch, 1958,. *Fall of the Roman Republic.* Penguin Books : Harmondsworth, Middlesex, UK.

Smith, A., 1776,. *The Wealth of Nations.* Current edition : 1970, edited by A. Skinner. Penguin Books : Harmondsworth, Middlesex, UK.

Soja, E. W., and Allan J. Scott, eds. 1986,. Special Issue on Los Angeles : Capital of the Late Twentieth Century. *Environment and Planning D : Space and Society,* 4/3, pp. 249-390.

Sweezy, P., 1950,. 'The Transition from Feudalism to Capitalism.' *Science and Society,* 14/2, pp. 134-157.

Sweezy, P., et al. 1954,. *The Transition from Feudalism to Capitalism : A Symposium.* Science and Society : New York.

Veblen, T., 1904,. *The Theory of the Business Enterprise.* Charles Scribner's Sons : New York.

Von Thunen, J.H., 1826,. *Der Isolierte Staat in Beziehung auf Landwirtshaft and Nationalökonomie.* Rostock.

Weaver, C., 1985,. 'Romance of the Roses : A Discourse on the Path to Regional Reconstruction.' *Environment and Planning D : Society and Space,* 3/2, pp. 239-258.

Webber, M. M., 1963,. 'Order in Diversity : Community Without Propinquity,' in *Cities and Space : The Future Use of Urban Land.* Ed. L. Wingo, Jr.,. Published for Resources for the Future. Johns Hopkins University Press : Baltimore.

Clyde Weaver
GSPIA, University of Pittsburgh
3N 22 Forbes Quadrangle
Pittsburgh PA 15260
USA

9. Space and creativity.
«Belle Epoque» Paris : Genesis of a World-class artistic centre

Paul Claval

By the end of the nineteenth century, artistic life was developing rapidly in the big continental European metropoles - Vienna, Berlin and Rome. Paris, however, was a centre of innovation both in its function as an artistic market and as a place of residence for sculptors, painters and other artists. From the 1890s to 1914, there were changes - Vienna, for example, grew in importance - but Paris remained the major artistic centre in the World. In the Interwar period, it kept this leading position.

There were long periods during which the French government developed an active artistic and cultural policy : this was the case during the *Ancien Régime,* from François I on, and again during the 1st and 2nd Empires. The tradition was resumed later during the 5th Republic, as a consequence of André Malraux's activities during Charles de Gaulle's presidency. The «BeauxArts» system which had been erected since the mid XVIIth century certainly played a role in the formation of *Belle Epoque* Paris, but the explanation for many aspects of its existence have to be sought elsewhere - in the geography and sociology of artistic creation.

1 The royal and imperial inheritance

The formation of the official «Beaux-Arts» tradition

François I was the first king to consider artistic and cultural policy a major aspect of his activity. During the Middle Ages, justice, faith and military ability were considered the principal royal virtues. A king's ability to write and recite

133

G. B. Benko and U. Strohmayer (eds.), Geography, History and Social Sciences, 133–142.

poetry, to sing and to dance was important, but his glory did not rest in his direction of the artistic development of the Nation. François I was the first king to demonstrate the new Renaissance interest in the Fine Arts and classical literature. By the early seventeenth century, such concerns were of paramount importance to the kings of France and their ministers (Richelieu and Colbert).

The basis of the «Beaux-Arts» system was conceived and began to operate between 1640 and 1700 (Picon, 1988a). Its main features were simple :

i) Artistic development was a reflection of the glory of the king, who has to develop an artistic policy, and to demonstrate his taste and his role in promoting new styles and new forms of art.

ii) The official policy benefited mainly Paris. Although many palaces, parks and princely residences were located elsewhere, the majority were within one hundred kilometers of Paris, which meant that architects and most craftsmen lived in Paris, and that paintings, sculptures and furniture were ordered from Parisian workshops or studios. Since a good part of the elite lived in Paris even after the migration of the Court to Versailles, wealthy mansions and *hôtels* employed many people on a permanent basis.

iii) Cultural policy was perceived as a means of improving the image of the Sovereign both in the country and abroad. This meant that his collections, his paintings, his sculptures, his palaces and his gardens, were open, effectively, to everyone. The Royal collections were essentially the equivalent of modern museums. The post-Revolution French museums were inheritors of the Royal collections. As a result, official policy offered models for young artists to copy in addition to new facilities for the development of artistic experience among the public.

iv) Cultural policy implied the development of a new type of institution : the Academy. Its role was to define sound criteria with which to judge paintings, buildings, sculptures, or plays. Its brief also included the organization and training of young artists (Picon, 1988b). As a result, Paris became the first city in the Western World with a coherent system of artistic training.

v) As the King and the Princes were the main customers, official direction played a decisive role in the orientation of artistic production.

The Napoleonic reorganization of the Beaux-Arts system

The French Revolution destroyed the fine arts institutions of the *Ancien Régime*, its academies and the teaching system linked to them. Napoleon had no real interest in art, and *Impératrice* Joséphine's interest extended only to her involvement in gardening - her park in Malmaison was a major achievement in landscape architecture (d'Arneville, 1891). But Napoleon was aware of the authority French Kings derived from their artistic and cultural policies. As a result, he decided to rescue the administration of fine arts and at the same time to rationalize it. The academies were resurrected and united in the *Institut de France*. Museums were created and artistic training was resumed. For the most part, the old structure was rebuilt and even became more efficient. Studies by Drexler (1977) and Middleton (1982) show the performances of the *Ecole des*

Beaux-Arts from the beginning of the XIXth century on. Its influence was decisive in France, but also abroad at least for the first decades of the XIXth century.

In emulation of his royal predecessors, Napoléon decided to promote a new style for furniture and decoration. But his influence was less decisive in the modeling to the urban landscape, particularly in Paris. Wars did not leave enough money to launch ambitious programs, to widen streets, to draw new avenues and to erect new palaces.

The Napoleonic organization was accepted by the Kings of France under the Restauration, but their attitude was utterly different from Napoleon's. They did not try to control the artistic policy of the fine arts establishment for which they were paying. Napoléon III was more ambitious, and through Haussmann, he launched a gigantic program for the modernization of his capital city (Saalman, 1971). Artistic aspects were important for him - the *Opéra* still bears witness to his ambitions.

The Fall of the Empire in 1870 marked the dawn of a new era. Republicans had struggled against the expanses of the Imperial artistic and cultural program - they had strongly criticized the *Comptes fantastiques d'Haussmann*. Under the 3rd republic system, the President had no real power and the government deemed it wiser to refrain from any directive intervention in the arts.

As a result, under the 3rd Republic, although there was an official arts establishment, with academies and teaching institutions more or less linked to them, there was no attempt to develop an official arts policy. Real power devolved to Academic circles but, as they were not backed by political authority, academies had hardly any control over new orientations. Arts policy was conceived of by the majority of politicians as an institutional inheritance to be left unchanged and certainly not to be used as an instrument for the glorification of the government. Under the 3rd Republic, a real separation of powers occurred : the executive ceased to elaborate artistic norms and restricted its role to financial support of the academies.

The royal and imperial inheritance was by no means negligible, as it constituted a performing fine arts training system. Official orders were significant enough to provide many artists with modest and sometimes substantial incomes. By the mid-nineteenth century, Paris possessed the largest concentration of painters, sculptors, actors and opera singers in the World. Certainly this was one of the reasons for the international influence of the city.

2 Components of artistic centres : social and geographic aspects

From royal patronage to market conditions

Until the end of the seventeenth century, official orders dominated the artistic market of a country like France. The King was not the only individual to buy paintings, furniture or sculpture, but the dynamics of the Court, so characteristic of the French society (Elias, 1973), resulted in the acceptance of Royal norms by all the courtiers. During the eighteenth century, although social and economic conditions began to alter, Versailles remained a model even for those unconnected to the Court. This influence was critical to the formation of the French taste, and played a role in the rising quality of craftsmanship so evident in Paris at that time.

With the development of a larger elite, conditions in the art world began to change. Royal patronage ceased to dominate the art market. Demand from more diverse and more independent individuals - bankers, merchants, lawyers, solicitors and that part of the aristocracy that lived in Paris and was more and more involved in business, and less and less in the Court - increased.

The Revolution accelerated the evolution : official orders declined during the Revolution, and never resumed their former level. But elites fled from Paris for only a short period and Paris remained one of the most fashionable of residences both for the French and for the international leisure class. Until the Second Empire, foreigners remained a small group. With Haussmann's programs, the situation changed. Paris offered a modernized townscape, the kind of environment suitable for those seeking beautiful apartments, harmonious perspectives, the amenities needed by upper class social life - theatres, operas, circuses, big cafés, restaurants - and a diversity of luxury goods and services not to be found elsewhere.

In a way, the best contribution of the Second Empire to the artistic influence of Paris, at the end to the nineteenth century, was Haussmann's variety of urban planning (Evenson, 1979) : it was responsible for the enlargement of the upper class living in or visiting Paris. Soon after the Franco-Prussian war and the Commune, the social scene grew calm and the wealthy congregated in Paris to buy garments, jewels, furniture or paintings, to attend theaters and to enjoy the beautiful women. The demand for artistic production grew rapidly.

Salons, art galleries and merchants

For a long time, the art market was organized around a central exhibition, the *salon*. It was an old institution. Colbert organized the First exhibition of the Royal Academy of Painting and Sculpture in 1667. This was the first organized artistic public show in the Western World - the first institution designed to diffuse information on Art production. During the *Ancien Regime*, the annual - or biannual - exhibition was restricted to the members of the Academy. It took the name of *salon* as it was held in the *Salon Carré* of the Louvre Palace, thus demonstrating the close links between the institution and the King. The French Revolution opened the *Salon* to artists who were not members of the *Académie des Beaux-Arts*. The Academy was in charge of organizing the exhibition including the selection of paintings, and as a consequence, the market may be still considered to have been a directed one. Potential customers felt secure in the knowledge that «bad» works had been refused. The *salon* was particularly important to young painters, as it offered them the possibility of discovery. Catalogs were produced from the eighteenth century, which extended the influence of the exhibition and critics began to cover the exhibitions from the beginning of the same century. With Diderot, the *Salon* review became a major literary activity : Theophile Gautier, Mérimée and Charles Baudelaire were famous for their critiques.

The official *Salon*, with its selection procedure, rapidly came under criticism and parallel unofficial *salons* were organized from the eighteenth century. For a long time, however, their significance was limited. Their popularity grew in the 1860s, particularly with the *Salon des refusés* (1864).

Until the middle of the nineteenth century, art merchants and art Galleries were relatively scarce, and their role was not paramount. By the 1860s, demand was more diverse than it had been twenty years earlier, and by the end of the century art works were moving more and more quickly. There were many reasons for this evolution : the influence of foreign customers, the development of Paris as a centre of counter-culture, the discovery of foreign, non-Western artistic traditions, particularly that of the Japanese in the 1870s, and the African artistic tradition thirty years later.

In Western societies, there was little room for counter-cultures as long as social organization was based on an established Church. Britain and protestant countries experienced greater diversity from the eighteenth century, but religious tolerance did not preclude a harsh collective control of morality and behaviour as well as of artistic taste.

Large cities played a crucial role in breaking with this kind of monolithic social organization. London and Paris offered the best opportunities from the eighteenth century; Vienna and Berlin from the mid-nineteenth century. In Paris, a tradition of political counter-culture had flourished at the time of the Enlightenment and never entirely disappeared. If the majority of the upper classes adhered to conservative ideologies, there was nevertheless a measure of diversity, and a will to explore new avenues in all aspects of life. By the Second Empire, with the rise of the middle classes and with their entry into the power and economic elites, the authority of the Academies began to decline.

The *Salon* ceased to be the key organizing factor in the field of artistic life. New institutions and organizations came to replace them, notably merchants and dealers which grew substantially in number during and the 1870s. There were many of them in Paris and their role was essential in connecting artists with provincial customers or foreigners. They were not however instrumental in launching young artists. Theodore Zeldin (1979: 11) notes that there were 104 such merchants in 1864. With people like Ambroise Vollard or P. Durand-Ruel and later D.H. Kahnweiler, the perspective changed: this new generation of art merchants considered it its responsibility to discover young artists, to finance them for either short or sometimes long periods, and to promote their works (Rewald, 1986).

A new art market geography appeared. Traditionally, the essential links had been with the Court - hence the role of the Louvre Palace - and with the *Académie des Beaux-Arts* - hence the importance of the Left Bank. Art galleries were originally located in these parts of the city, but an evolution soon occurred and the art market moved towards its customers. By the end of the nineteenth century and the beginning of the twentieth, the double polarity of the Art market characteristic of Paris for about 80 years was clear : *rue de Seine* on one side, *rue Saint-Honoré* or *rue de la Boëtie* on the other.

Artist and their colonies

The third element in the modern art market system were the artists. From the seventeenth century Paris had been home to many of them and Theodore Zeldin (1979: 113) notes that in 1885 «3851 artists proposed their works to the Salon». Two factors were responsible for this :
i) Paris offered the best *Beaux-Arts* education in the World from the beginning of the eighteenth century to the end of the nineteenth century. Royal collections, and later museums, facilitated the training of young artists. Through copying the great masters they were able to emulate their techniques and their styles. The workshops and ateliers which began to develop in the 1720s and 1730s and which brought together young artists made them work in competitive-cooperative communities for extended periods. They were directed by the best artists - or some of the best artists - of the time.
ii) The sheer size of the market, the fame of the *Salons* and the role of critics were also important. From the 1860s foreigners began to consider Paris a necessary stage in their formation, and increasingly moved to and settled there.
After their apprenticeship, painters could leave Paris. It was important for artists to be able to meet other artists and to compare techniques and works. An ambiance conducive to creative work required that painters congregate in colonies - but the colonies needed not necessarily be urban. After 1850, with the advent of the railways, more and more painters left Paris and settled for long periods in rural areas, in small cities, in sea resorts : Barbizon, Argenteuil, Pontoise, Honfleur, Pont-Aven, and Provence attracted colonies of painters. With impressionism, contact with nature assumed a new importance (Brettell and Schaefer, 1985). Notwithstanding, time in Paris was necessary to maintain contact with merchants and critics, and to allow stimulation from a wider array of artistic experiences.
In Paris, artists' problems were numerous. This hardship was a consequence of the new economic independence of artists. Traditionally, painters and sculptors had been housed by their patrons. In a market system, their room and board disappeared. For the majority of young talented people, it was difficult to pay the high studios rents in the centre of Paris and apartments and food were also too expensive in the bourgeois Western part of the city. Summering in small villages was a good solution. Winter was the time to be back in the city. Artists enjoyed the popular atmosphere of many low income areas - but it was important for them to develop links with the demand side of the market. Often they chose to settle along the border between low income and high income Paris, far from the central part of the city, with the exception of the decaying area of *Marais* (Sutcliffe, 1970). Manet left the *Café de Bade*, in the centre, for the *Café Guerbois,* avenue de Clichy, in 1866. Ten years later, many artists met at *Café de la Nouvelle Athènes,* place Pigalle (Rewald, 1986: 133, 253). Hence the success

of Montmartre at the end of the century, and of Montparnasse, in a symmetrical position to the South, before and after World War I.

At a time when new directions had to be explored the poorer parts of Paris were also the main focuses of counter-culture (Chevalier, 1958). These areas attracted both artists and counter-cultures leaders. They frequented the same *cafés* and both groups throve on the marginal atmosphere of these parts of the city, where they found inspiration and support.

The components necessary for a large art market were present in Paris at the end of the nineteenth century : i) a strong demand, with an appreciation of new ideas and new experiences, and strong connections all over the World; ii) a commercial system, with *salons* to give opportunities to new talents, and merchants to prospect new curiosities and support young artists; iii) numerous well-trained artists with a fine tradition of workmanship, and proximity to potential customers, merchants, and each other.

3 The dynamics of success

Paris was one of the main centres of Western artistic life from the end of the seventeenth century, and it was dominant by the end of the eighteenth century. The French Revolution, the First Empire and the Restauration did not interrupt artistic activity, but Paris' international influence was certainly diminished by political events. Recovery came with the July Monarchy and the Second Empire. Paris had long been a centre for the arts, literature or fashion, but its attractiveness was substantially improved by events such as the international exhibition of 1855, and by the new townscape created by Haussmann's urbanism. Paris became the place where new forms of social life for the well-to-do or for intellectuals were devised. The number of foreigners who chose to stay in Paris for short or long periods grew rapidly. Up to the July Monarchy, many people had come to Paris from Germany and Central Europe. With the economic take-off of this part of Europe and with Franco-Prussian rivalry, immigration from the Germanic countries declined, with the important exception of the Jews. Poles and Russian were attracted from early in the nineteenth century, and continued to move to the French capital. But what was new was the growing popularity of Paris among British or Americans. For many Americans visiting Europe, Britain was the ancestral home. Germany was considered the source of new techniques and first-rate Universities. Italy offered its ruins, its renaissance palaces or churches and its museums. But France, and more specifically Paris, offered excitement, the amenities of daily life and new forms of art. Many Spanish and Latin-Americans also stayed in Paris for extended periods. Haussmann had set the stage for Paris' artistic dominance.

A comparison of *fin de siècle* Paris and Vienna (Schorkse, 1983) suggests that esthetics were less central to the reflections of the French upper and middle classes that to their counterparts in Austria. In France political life and social problems were considered more significant. This was one of the reasons that the

French government ceased to control artistic life. Academic structures remained strong, but had lost their political support. They were effective in making Paris attractive to young artists - but they proved unable to prevent the rise of new artistic conceptions.

Paris was a city of new-comers and self-made men : economic life had been developed in France by people who did not come from the traditional aristocracy; the landed gentry had moved back to its provincial estates at the time of Louis-Philippe. The State had still slight influence on economic life, and civil servants could not easily move to the private firms. Thus, the new business class was composed either of self-made men, or of engineers from the new *Grandes Ecoles*, where they were provided with an excellent technical training, and imbued with the idea that the analytical method that they had learnt could be effectively applied in every field - in the arts as well as in literature. The attitudes of this new elite differed from those of the traditional elites. They had less respect for Academic formulas. They did not wish to live in the presence of mythological scenes - they knew about Antiquity, but considered its old tales such as Offenbach's La Belle Hélène mere diversion. The new élite was without metaphysical or religious angst. It despised the old image of culture. As a result, it was open to new forms of art and sought in its art pleasant scenes and beautiful things rather than intellectual experiences.

The crisis of the old artistic establishment was evident from midcentury : Courbet presented his paintings independently of the official *Salon* during the International Exhibition of 1855. The *Salon des refusés* appeared in 1864. *Salons* became more and more numerous toward the end on the nineteenth century. At the same time, art galleries began to play a new role in promoting young painters and uncovering foreign customers - Russian merchants, American businessmen, or their wives. For such buyers, the absence of references to traditional aristocratic culture in French painting and sculpture was significant as it allowed them to appreciate what they bought without effort.

The dispersal of artistic colonies all over France began in the late 1850s and the early 1860s as a result of the new railway lines : it was easier to spend a few months in a pleasant rural environment and to be back in Paris during the autumn and winter seasons or whenever it was necessary to meet potential customers. As a result, although artistic life appeared decentralized and while artists lived approximately half the year out of Paris, their links with the Parisian market were strengthening. They were still significant provincial art markets - in Lyon, in Nancy, in Rouen etc... - but the role of Paris was growing.

During the 1860s and 1870s, Paris had a troubled political life, with much social unrest. Many artists were left-wing militants - Courbet, during the Commune, destroyed the Vendôme column and later had to pay to have it rebuilt. But the majority of artists did not feel that they had to depict factories and workers in order to promote their ideals. In this, they differed from the novelists such as Zola. They were eager to experiment with new ideas and they were open to many counter culture themes, but they did not think of themselves as social militants. They considered that their true responsibility was to explore the world of sensations and to break down age-honoured artistic conventions.

Conditions changed at the end of the nineteenth century. The influence of Academic authority had by then considerably declined. The balance between the official *Salon* and private *salons* or art galleries, which was characteristic of the 1870s and early 1880s, disappeared. The artistic market was increasingly competitive. As a result, critics became more powerful : traditionally, they had judged painting and the skill of painters. From the late nineteenth century, they began to focus on the novelty of the artistic conception. Avant-garde art was born. The evolution started with the creation of the *Salon des indépendants* in 1884 : Seurat, Signac, Toulouse-Lautrec and Van Gogh were discovered through it. In the early twentieth century, the evolution accelerated with the *Salon d'Automne* and the promotion of *Fauves*.

Paris was the first city to develop an avant-garde art market : it was the natural outcome of a competitive system in which the only accepted esthetic value was novelty. The evolution was paralleled in Vienna, where such a market appeared just before the First World War - but the Austrian defeat shattered the evolution of the Viennese art world. Similar developments took place in New York only during the Second World War and as a consequence of the isolation of the Parisian market (Guilbaut, 1988).

As a result, for about 50 years, Paris found itself exclusively capable of launching new art forms in the Western World : artists felt obliged to move to Paris because only there could they experience the kind of competitive atmosphere necessary to the avant-garde, together with the market conditions essential to the promotion of their art.

After its initiation, the movement developed according to its own logic. The avant-garde mentality was born of the criticism of the formulas of official art at a time when people had grown tired of it. By the 1930s, avant-garde art had lost touch with popular sensibility. It had once again become elite. Paris' immanent crisis as an artistic centre was also linked to the rise of competing centres - essentially New York - and to the ivory tower mentality of the younger avant-garde artists. The crisis was postponed by structural changes occurring in the demand for art - the growing significance of company patronage. But the difficulties were multiplying, and few were able to understand their causes.

The history of the international influence of *Belle Epoque* Paris is fascinating : historical inheritance was paramount in the development of a milieu of artists and merchants, in the formation of taste and in the establishment of norms, but influence grew mainly precisely when State intervention was dwindling and when new mechanisms of promotion were appearing. These new mechanisms relied on a decentralization-centralization dialectic : artists spent more and more time in remote rural areas, but at the same time, the function of Paris became more central in the demand-supply confrontation. Once the new mechanisms had appeared, the accumulation process worked according to its own inner dynamics for half a century. After World War II, the situation changed : Paris lost its monopoly, and its artistic system lost credibility.

Note :

Text re-vamped by Anne Godlewska.

References

Arneville, M. B. d', 1981, *Parcs et jardins sous le Premier Empire*, Paris: Tallandier

Brettell, R. and Schaefer, S., eds., 1985, *L'impressionnisme et le paysage français,* Paris: Ministère de la Culture

Chevalier, L., 1958, *Classes laborieuses, classes dangereuses*, Paris: Plon, new ed., L.G.F., 1978

Drexler, A. ed., 1977, *The architecture of the Ecole des Beaux-Arts,* New York: The Museums of Modern Art

Elias, N., 1973, *La civilisation des moeurs*, Paris, Calmann-Lévy: 1st German ed., 1939

Evenson, N., 1979, *Paris: a century of change, 1878-1978*, New Haven: Yale University Press

Guibaut, S., 1988, *Comment New York vola l'idée d'Art Moderne*, Nîmes: Jacqueline Chambo; Am. 1st ed., Chicago, The University of Chicago Press, 1983

Middleton, R., ed. 1982, *The Beaux-Arts and nineteenth century French architecture,* London: Thames and Hudson

Picon, A., 1988 a, *Claude Perrault ou la curiosité classique*, Paris: Picard

Picon, A., 1988 b, *Architecte et ingénieurs au siècle des Lumières*, Marseille: Parenthèses

Rewald, J., *Histoire de l'impressionnisme,* Paris, Albin Michel ; 1st Am. ed., 1946

Saalman, H., 1971, *Haussmann: Paris transformed*, New York: George Braziller

Schorske, C. E., 1983, *Vienne fin de siècle*, Paris, Seuil: 1st Am. ed., 1961

Sutcliffe, A., 1970, *The autumn of central Paris*, London: Arnold

Zeldin, T., 1979, *Histoire des passions françaises, vol. 3, Goût et corruption*, Paris: Recherches; 1st Brit. ed., 1973-1977

Paul Claval
Université de Paris IV – Sorbonne 'Espace et Culture'
191, rue Saint-Jaques
75005 Paris
France

10. From Weimar to Nuremberg :
Social legitimacy as a spatial process in Germany, *1923-1938.*

Ulf Strohmayer

<blockquote>
«In every era the attempt must be made anew to wrest tradition away from a conformism that is about to overpower it.»

Walter Benjamin
Theses on the Philosophy of History
</blockquote>

The question of National Socialism, then, here and today: what does it mean to speak of *it* in the singular ? Indeed, what does it mean to speak *of* it at all ? Almost fifty years after the liberation of Germany, in fact Europe, from a most ruthlessly effective totalitarian political system, the problems posed by the historical emergence of *something like* National Socialism remain manifold. Its causes, we can read in an ever growing amount of literature, are many and different, converging, as it were, in the concrete, if at times conflicting conditions that made «it» possible. «It» *was*: twelve years in German history in dire need of interpretative efforts; efforts, or so an implicit hope would make us seek, which eventually and with hindsight will uncover *what* exactly these twelve years exemplify. Do they represent a «totalitarian dictatorship», as liberals tend to believe ? The most gruesome and efficient of European «fascisms» ? A consequence of capitalism in a time of crisis ?

In this essay, I want to pursue a different mode of reasoning. Rather than taking the distinct historical formation of National Socialist Germany for granted, only causally to link this particular political entity with prior events and forces decisive within German history at large, I will look at material manifestations of the time in question in order to construct, as it were, conditions of possibility *from within*. This internal mode of historical analysis, as I hope to demonstrate here, allows us to approach an understanding of historical concreteness without

143

G. B. Benko and U. Strohmayer (eds.), Geography, History and Social Sciences, 143–168.
© 1995 *Kluwer Academic Publishers. Printed in the Netherlands.*

relying on any of the possibly flawed assurances of knowledge on hindsight. «Constructing history from within» is of course to beg the question, for whatever «history» is, for us, here and today, it does not take after the tangible notion of a doughnut: its «inside» cannot but become a function of prior choices on the side of an historically interested scholar. Lacking any interest in masking these choices, this essay is consequently about the necessity, the impossibility, and the promise of a *frame*. For a promise it is from the beginning (which for you, I suspect, came about by either glancing at the table of contents of this book or else through some other kind of external reference): «Germany, *1923-1938*», «legitimation», «modernity», and «public life». A promise and a frame, a frame in space and in time, which is itself and in turn framed thematically. And indeed, we might ask, is such not the general structure of *any* promise, for what would there be to promise *without* a frame, *without* frames delimiting possibility ? Even knowing, as we do today, that no picture fits a frame unless it has been commissioned for this fit in the first place, *«from the beginning»*, a promise still is promise, a kind of contract between you and me. Why note the obvious ? Because the title of what is to come, which deliberately neglects to mention National Socialism by its proper name *while* promising to talk *of*, and thus include «it» into *its* spatio-temporal frame, is precisely about «the obvious», and about why we should resist it. Unable to circumvent the economy of a promise, yet at the same time incapable of fulfilling it, my goal in this essay cannot but be to open up spaces, to speculate in the best, if theoretically de-legitimized sense of the word, and, in so doing, to offer a gift instead. And proverbial wisdom notwithstanding, one should always look a gift-horse in the mouth.

1

The more so, if a specific gift reveals itself as a Trojan horse right away. «Germany, *1923-1938*», a frame to be sure, but a frame that encompasses and thereby mingles two distinct generalities, two separate entities: the «Weimar Republic» up until 1933 and Nazi Germany thereafter. No doubt, the former started some five years before 1923 in the aftermath of defeat, and the worst was yet to come in 1938, but the attribution of proper names, the *differentiating* framing of events around the «seizure of power» (the «*Machtergreifung*») by Adolf Hitler and his followers on 30 January 1933 in Berlin allows us - quite literally - to come to terms with the sheer monstrosity of history, of German history, that is. All that we need, this differentiation suggests, is a beginning and an end and the in-between becomes, well, «itself». Thus framed, a historical entity can now be causally connected with other entities and from hence be contextually explained. How else could we separate «democracies» from «dictatorships» ? And yet, in separating and thus periodizing, do we thereby understand this difference any better ? An apparent transparency of historical categorizations notwithstanding, allow me to demonstrate how a reframing of obvious and justified placements of events can help us to see «difference» in a new light. For to this day, I will argue, our various interests in the perspicuity of historical differentiations not only obstruct, but actively prevent a particular kind of insight about which I will have to say more later on in this essay. To this very day, we identify by framing, we name epochs and specify spaces, and in so doing ban the ghosts of the past only to deprive ourselves of demanding possibilities to understand differently what we need to understand most importantly, namely the banal workings of societal legitimation *as it unfolds* within the structure of everyday, public life. But if this «banality», which so pointedly has been brought to our attention by Hannah Arendt (1963), threatens traditional «knowledge on hindsight», it likewise hosts the possibility of a comprehension *beyond an impossible justification*. The problems are numerous, for

> (w)hatever its other aspects, the everyday has this essential trait: it allows no hold. It escapes. It belongs to insignificance, and the insignificance is without truth, without reality, without secret, but is perhaps also the site of all possible signification (Blanchot, 1987: 14).

To arrive at a point where we are able to see this possible signification, let us first take a look at where we are coming from. (For even a new kind of insight requires a frame from which to depart).

2

Which is, finally, to remind us that every history has a geography, *one* at least. And within the space of a differentiating, generalizing frame of historical entities, this geography is quite often condensed into a quite persistent form of «icon» or «spatial cliché», it becomes exemplary. The spatio-temporal frame at hand is not an exception. In fact, if ever there have been «spatial clichés» expressing historical entities to create difference, it is within the frame of our present encounter. On the one side and before «1933», the «modern», «roaring» and violent «twenties» create unintelligible and confusing, yet at the same time pluralistic spaces within almost every facet of public life; on the other side and after «1933», a repressive, classical linearity leaves its imprint on a truly archaic landscape of increasing political silence or public uniformity. One cliché merges Weimar with Dada, while another links Nuremberg to its rallies and to the unity of *Volk* and space. To deny the validity of the resulting *exemplary* images is to scoff at the millions that paid the price for this difference with their lives, their sanity, and, yes, *their* «banal» normality. And yet, to understand this difference (or the conditions of possibility of any such marginalization in general), our comprehension will have to disregard what it seeks to explain and go against the grain if it is to escape the circularity of conventional, reassuring wisdom.

3

If this formulation has itself all the air of a paradox, it also once again embodies the uniqueness of geography. Which is to say that being geographers ourselves, we are in a position *sans pareil* to contribute to this «understanding against the grain», because for us «frames» have always been relative means to an end at most. For us, any single «frame» constantly embodied the possibility of difference: we have always known (if at times forgotten) that a change in *scale can always* engender a different picture. But this mutual relativization of different frames is enacted on more than one analytic stage. While historians have at best conceded a *diachronic* relativism of general frames (the development of capitalist economies should suffice by way of example here), it is time finally to supplement this with a *synchronic*, and thus inherently geographic relativism of sorts. *Between frames*, there is not simply random noise, but uncharted terrain, the hitherto uncontextualized realm of spatial non-synchronicity. Within this realm, the dreaded ghost of «relativism», postmodern or otherwise, need not be a threat, but could become a possibility instead, a possibility to understand historical change better. What I have in mind, of course, is hardly revolutionary. The history of modernity, if indeed there is such a thing, is distinguished in a *temporal* sense by successive sequences of innovations within an expanding capitalist mode of production. But change, as we all know well, is often a gradual process, leaving the belated analyst with a synchronicity of difference *at any given time*. As Ernst Bloch, to whom these pages owe more than a cursory debt, once phrased it:

> Not all people exist in the same Now. They do so only externally, through the fact that they can be seen today. But they are not thereby living at the same time with the others. ... Various years in general beat in the one which is just being counted and prevails. Nor do they flourish in obscurity as in the past, but they contradict the Now; very strangely, crookedly, from behind (1991: 97).

Today's societies of mass consumption, for example, coincided (if by no means always peacefully) for the longer part of their «history» with bourgeois worlds of trade and commerce, thus rendering their particular historical frame, their general delimitation, so delicate an issue. But «any given time» comprises any given number of places along an urban-rural continuum or hierarchy, places which add their share of non-synchronicity to any notion of historical change. Most of us engaged in historical research will have experienced such spatial non-synchronicity in the form of «local» (or «particular») relativizations of general frames or historical «trends». And yet, in so juxtaposing generals and particulars, we perpetuate the dominance of «time» over against «space»[1] *when we could take advantage of this very relativization instead*. For it is here, *in* the inherent spatiality of any historical event, that historical tensions originate and become manifest, open to analytical experience.

4

For what, we should consequently ask, *is* a «particular» ? Since the relationship between generalities and particularities, touched on in the above section of this essay, remains at the core of any social scientific endeavour, allow me to clarify some rather important issues. Right from the beginning of this endeavour, sometime in the 19th century, with the institutionalization of possibilities to justify synchronic analyses of different historical entities in general, «the particular» became *the thing to be accounted for*, unless «it» fit the general description of an epoch at hand in all of its major synchronic aspects. Unless, that is to say, «it» was «typical» (or «exemplary») anyway. And account we did, only successfully to do away with whatever particularity was left at the beginning. In other words: until a particular had become a generality by its own right.

Lately, with the emergence of a «new» regional geography and various attempts to synthesize «generals» and «particulars» under the heading of some realism or another, the said dichotomy has been subjected to increasingly skeptical questioning and consequently been loosened somewhat. Necessary as this skeptical intervention remains, most often does it continue to express more of a desire to square the circle of understanding justifiably than articulate a promising path toward a future. Because even here, in the *substantial* blurring of divides, does the *formal* possibility (and hence *academic* profitability) *of* analytic distinctions *in general* remain largely taken for granted. How else are we to read theoretical justifications for coining a term like «locale» ? The problem is that *any* term conceived as a tool to synthesize both general and particular, has to rely on always already existing analytical distinctions like our present periodization, which thus inevitably form the backcloth for further systematic inquiries. A historical geography of the rise of Nazism to power, for example, would still take «Nazism» for granted and thus, as I hope to demonstrate soon, neglect to get us anywhere near the concrete conditions of possibility *of* Nazism. But how is one to avoid this trap, if indeed a trap it is ? By not insisting on some *a priori* standard of comparability between different historical evidences. In other words, by not asserting the (presumably neutral) existence of a historical yardstick, measured against which historical events could become relative *to something non-relative* in the first place.

Rather ensuingly, the «particular» evidence presented here will *not* dwell on the justification of standards for comparison. Not that they are lacking. On the contrary, but has that really ever been an issue «with the benefit of hindsight» ? Collecting material remnants (or «ruins», to use Walter Benjamin's term) from the arbitrarily chosen «provincial» town of Freiburg im Breisgau, I consequently desired to unlearn the latter-day fact that both this mid-size, Catholic and conservative university town far and away from Berlin (*1925*: 90.475 inhabitants; *1939*: 110.156 inhabitants; cf. Quilitzsch, 1968: iii) and the surrounding *Land* of Baden, being neither «Nazi-strongholds» nor islands of working class dominance, reflect a sense of (bourgeois) «averageness» that becomes manifest *inter alia* in a documentation of national election results up until the *Machtergreifung* and its legitimating elections in March 1933 (see Tables 10. 1 and 10. 2 and Figure 10. 1).

Table 10. 1: National Election Results in the German Reich (in % of total valid votes)

	1924 May	1928 May	1930 Sept	1932 July	1932 Nov	1933 March
NSDAP	3.0	2.6	18.3	37.1	33.0	44.2
DNVP	20.4	14.3	7.1	6.0	8.5	7.9
DVP	9.9	8.6	4.6	1.0	2.0	1.0
Z	17.5	15.2	14.8	15.7	15.1	14.0
DDP	6.3	5.0	3.7	1.1	0.9	0.8
SPD	26.0	29.8	24.6	21.6	20.3	18.4
KPD	9.2	10.7	13.1	14.3	17.0	12.2
others	7.3	13.0	13.8	3.0	3.5	1.5

source: Abraham (1989, 43). *note*: see Table 10. 2.

Table 10. 2: National Election Results in Freiburg im Breisgau (in % of total valid votes)

	1924 May	1928 May	1930 Sept	1932 July	1932 Nov	1933 March
NSDAP	—	1.3	14.1	29.8	22.8	35.9
DNVP	10.3	8.4	3.8	5.7	9.9	8.3
DVP	11.0	10.4	12.0	1.6	3.6	1.6
Z	34.5	32.7	30.7	34.7	32.1	29.5
DDP	8.5	5.9	—	2.1	—	—
SPD	14.3	26.6	21.6	15.6	15.7	14.1
KPD	9.5	3.9	5.9	7.8	12.2	8.0
others	11.9	10.7	12.2	3.0	3.8	2.7

source: *Freiburger Zeitung* 5 May 1924, *Breisgauer Zeitung* 7 November 1932, 6 March 1933.
note: Abbreviations used in both tables: NSDAP= National Socialist German Workers Party; DNVP= German National People's party; DVP= German People's Party; Z= Catholic Center party; DDP= German Democratic Party; SPD= Social Democratic Party of Germany; KPD= German Communist Party. A number of smaller parties were represented in the German Reichstag and reflected a number of quite powerful, if regional political movements.

Also, I was not very much interested in the relational fact that the local mechanisms pertinent to the «seizure of power» in 1933 and its aftermath portray striking analogies with the larger trends within the Reich *in general* (as, for example, the rise of the NSDAP to power, documented in Grill, 1975; the political *Gleichschaltung* or «bringing into line», presented in Mittendorff, 1988, Bräunche *et. al.*, 1983, and Schnabel, 1983; or the gradual eradication of Jewish life, illuminated in Böhme and Haumann, 1989). History, and let me emphasize this, is not the result of a smooth translation of processes between scales; rather, it genuinely *is*, and *only* is «between scales», a *synchronic* and simultaneously *diachronic* relation of tensions.

5

It is consequently within a *relative* and spatially diverse frame that I wish to propose an elucidation of *Nazism* that does not rely on a prior fixation of general distinctions. If these distinctions, so appropriately expressed by the kind of «spatial clichés» I pointed out earlier, had to serve the understandable but misguiding interest of discriminating between democracy and its totalitarian «other», they also sheltered the flexibility and parasitic nature of *power* from our analytic capacities. But worse than that: they obscured the concrete mode of legitimation employed by National Socialism *as an employment of power* to gather a mass of outspoken or silent followers. For it is precisely in the uncharted territory between frames that the possibility of Nazism becomes the pervasive possibility of legitimizing any use of power as a *necessary* means of history. Contrary to common belief, the contents of this legitimation mattered little. What did matter then, and still matters today, is its effectiveness. And the «effectiveness» of National Socialism, if such an obscenity could be morally acceptable in view of the atrocities implied, resided expressly in a masterful use of spatial non-synchronicity. In other words, «history» herself, or, more to the point, «modernity» itself provided the means for a temporary legitimation of National Socialism[2].

6

By now, most of us are familiar with the following, diachronic analysis: the German Reich, a relative latecomer at the table of European nation states, had to digest the results of industrial and cultural modernizations at an unprecedented pace in history. As a direct result, accentuated furthermore by the defeat in World War I, «traditional» and «progressive» elements within society fought out the future of the Reich during the Weimar years, *and we all know which side was victorious in 1933*. But this conventional, static view of power is turned upside down once we de-frame the underlying diachronic approach and relativize it spatially. The resulting new frame renders it possible to see a «Weimar modernity» that was announced (Figures 10. 2 and 10. 3), uncovered as a strategically placed joke (Figure 10. 4), and which, in the end, never even touched ground to leave an imprint on the cultural fabric of «provincial» everyday-life (Figure 10. 5). Even mundane symbols of «progress», like the *Zeppelin* in this example remained just that: «symbols» with the inflated value of imported otherness. Within this new frame or new promise it also becomes possible to apprehend the use of this spatial non-synchronicity in the legitimation of National Socialism (Figure 10. 6), as in the legitimation of *other*, likewise flexible forms of power (Figure 10. 7). Strange impersonations of Lindbergh indeed! In other, potentially more familiar terms, far and away from Berlin, and closely tied into a «democratic» core-periphery relationship, industrial and cultural modernity were but a distant dream for a majority of Germans during the Weimar years, u-topias whose time had come *elsewhere*. Here, in the so called «provinces», the promise of National Socialism was the promise of a dissolution of historical tensions, a dissolution of modernity with modern means (Stern, 1984: esp. 14; also Nipperdey, 1980: 381). And again, what might on the surface appear to be yet another insoluble paradox, is in reality a rather mimetic insight *between frames*. Linking the periphery to the core (Figure 10. 8) by making the core part of the periphery (Figure 10. 9) and *simultaneously* identifying this «egality» with a political movement (Figure 10. 10): New Hampshire, once every four years, stills suffers from this decidedly modern employment of «propagandistic» means to an end in control. A remarkable, «modern» shrinkage of space *and its use* becomes visible: Hitler was the only politician of the Weimar Republic to take extensive advantage of the Aeroplane during election campaigns. The spatio-temporal distance «covered» on this local ad, i.e. between Freiburg (7.30 pm) and Radolfszell (9 pm) still was a distance of almost 6 hours by train in 1932. The message is clear: «Be part of progress, *no matter where you live*, and being part of progress signifies *belonging to us...*» (Figures 10. 11 and 10. 12). *One People, one Reich, one Führer*, and eventually one thousand years crushed into twelve. In other words, there is indeed «a logic of fascism» (Nancy and Lacoue-Labarthe, 1990: 294), and this «logic» is decidedly modern in nature (Figure 10. 13):

> If the tendency of modernism, from its roots in romanticism, was «to objectify the subjective», to translate into symbol subjective experience, Nazism took this tendency and turned it into a general philosophy of life and society (Ecksteins, 1989: 314).

7

And it is by no means merely a question of aeroplanes.[3] Starting in «1933», we can document an increasing saturation of everyday life *across Germany* with the products of a mass produced modernity, a permeation which (as we can now see more clearly) carried the stamp of National Socialism most effectively where this stamp remained invisible. Moviestars and the *Autobahn* alike united the country both metaphorically and for real, while a minority - and it is always «a minority» - was forced to pick up the tab of this nationwide rendez-vous with modernity. In light of *this* frame, Chaplin's *Modern Times* and Riefenstahl's *Triumph of Will*, both produced in 1936, truly were «about» the same phenomenon: the former in a critical manner, the latter jubilantly, but in either case, the vanishing point is in the machine. Only that Hitler, descending from the skies in the opening scene of *Triumph*, knew about the importance of space and knew how to take advantage of it *without* appropriating it openly. There is, in other words, absolutely no reason *not* to take literally his proclamation in 1936, issued at the opening of the annual Frankfurt auto fair, that «the German people have absolutely the same desire as the American» (as quoted in Schäfer, 1981: 119, my translation; rather fundamental here Reichel, 1991: 115-28). There is no reason, because it was through this identification, that National Socialism could outplay capitalism for its own means, all the while being constantly used by capitalism itself. Undifferentiated, both turned ubiquitous, only to become not unlike electricity (Figure 10. 14). Completely missing out on this decidedly *spatial* «package», oppositional arguments remained impotent. To quote the official Social Democratic newsletter from June 1937, published from exile in Prague:

> The German youth is not inwardly filled with the idea of National Socialism. Rather, there is enthusiasm for sports, technique ... The young German no longer wans to become an engineer on a train but wants to become a pilot instead (Deutschland Berichte, 1980: 843, my translation).

But in 1937, becoming a pilot and becoming German were one already, united under the «big tent» of a particular ideology.

8

> Paltry shops are bursting with pots, cheap clothes, rubbish from the big city; far too many tinned preserves are ageing between them. (...) Fewer and fewer characters permeate the small town, less and less language is still stewing in its own juice, less and less country loaves, good old days in the newspaper. Instead, yesterday's cliché rules, and just as the shops have their tinned preserves, public opinion comes ready set, freshly churned, as dross from Berlin. (Ernst Bloch, «Small Town», *Heritage of our Time*.)

9

The same modern «shrinkage» of space for the sake of control, while almost completely missing during the 1920's, can be observed almost everywhere in «provincial» everyday-life after 1933. «You really still don't have a radio» ? (Figure 10. 15) asks the local ad, «you're really still unconnected» ? (Figure 10. 16) is what it implies, and «you're really *still* not a Nazi» ? (Figure 10. 17) is what it means. State-sponsored tourism within the everpresent «Strength through Joy» («*Kraft durch Freude*») movement brought «stressed out» city folks in contact with nature anywhere in the now accessible Bavarian Forest or along the Baltic Sea, while simultaneously «shipping» (Figure 10. 18) people from the «provinces» to places as exotic as New York City. Even «time» itself had changed. Once, in 1927, measured by cuckoo clock even when a modern industrial fair was being advertised (Figure 10. 19), it was now a unifying time, racing across national divides under the banner of progress, barely disguising a clearly legible subtext (Figure 10. 20). As Adorno observed:

Fascism was the absolute sensation: in a statement at the time of the first pogroms, Goebbels boasted that at least the National Socialists were not boring (1987: 237).

Once again, we did not listen patiently enough, for National Socialism did not disguise its «modern» proclivity. In insisting time and again that it represented a «movement» («*Bewegung*»), rather than a «party», National Socialism broadcasted its *spatial* structure for everyone to know (Eichberg, 1989: 40-1). Everywhere and nowhere, it genuinely epitomized a sense of Nietzschean action, a forward motion that justified itself *by* itself. As Robert Musil put into the mouth of his «man without qualities» in the mid-1930's:
«I promise to you», replied Ulrich solemnly, «neither I nor anyone else knows what is true and what truth is, but I can assure you that it is about to be materialized» (1965: 135, my translation).

10

We may thus concur with Susan Buck-Morss when she insists that
fascism was not an alternative to commodity culture, but appropriated its most sophisticated techniques (1989: 309),

all the while asserting that this «appropriation» was a spatial one at bottom. What do we gain in so insisting? A more complex and thus more illuminating portrait of a totalitarian «normality» that thrived, at least publically, on the same modes of legitimation we still find in democracies today (Paul, 1990: esp. 12-3). If anything, this should caution us to *any* kind of «normality» within *any* kind of society, without providing us with any of the reassuring means that come with an *a priori* frame of identification. Lest we forget, 1938 closed on a night that had become a possible part of *some* «normality» during years of a spatially flexible employment of power (Figure 10. 21). Even a *Reichskristallnacht* can be made part of an acceptable price for progress. Allow me to quote from a letter written «the morning after»:
Father is now soon going to be finished with the drawings for Weber, maybe around mid-November. He's working like mad; today he came home from school already roundabout 10 a.m., because the synagogue burned and was later to be blown up. The people at the school over the road had to open all the windows, because of the pressure. As you can imagine, remaining in the hallways, the boys didn't really get much done. They were allowed to go home (Letter F.Rieß to R.Rieß, 10 November 1938. Municipal Archives Freiburg («Stadtarchiv Freiburg»), M 2-127a-12).

Lest we forget, not everyone was allowed to go home.

11

After that the doorbell rang at night. Two men came to pick up father. It was the ninth of November and father was sent to a concentration camp. The date, the word, and the abbreviation were not yet familiar ones, and it was only during hours, during days, that the word «pogrom» became attached to what had begun. The synagogue burned. The Jews themselves had set fire to it, the Gestapo men said in the night. That was clear and naked and utter nonsense, not even adorned with the usual effort of a justification or an excuse. It was the legitimation of the lie as a means of extermination. Whoever pronounced it knew that he lied; whoever listened to it or read it knew it was a lie, but this was irrelevant. Nobody thought of taking words literally any longer. Everybody understood: destruction.

Father was gone, disappearing without a trace. He could have been in town or thereabout. He could have been on a transport or dead. For the next couple of days, our house was filled with distraught and crying women, daughters, and sisters of men, fathers, and brothers that had been taken away. Banker Dorn's villa was deserted. Rumours were circulating, nourished by fear, and nourished by hope, but no one knew.

The city, inhabited for years and intimate, seemed nonetheless to be devoid of any secrecy. We knew the names of its streets, alleyways, squares, and all its official buildings. Our life was one with the work and the sleep of the city. We heard, saw, smelled, and observed everything like before. The music in the inns, the smells of bread, leek, and apples, and the stroke of the *Münster* bell were all unaltered. And yet, right in the centre of everything, there was a blind spot where father had disappeared.

Then we learned the word «Dachau». And that one was allowed to send warm clothes. From hence, the crying women met at the long baggage ramp of the freight depot. Two clerks took the boxes and suitcases and packets, all with the same address to many men with «Israel» as their first name. It seemed as if these clerks were messengers of some Hades, uniformed angels of the in-between, knowing perhaps, but never open to conversation.

One went home again, taking the long way across the bridge over the railroad tracks, passing through the *Markgrafenstraße* and across the *Münsterplatz* into the house. Everything seemed as ever. And whoever ignorant asked to see father, was told he was gone on a vacation.

The phrase «gone on vacation» has appeared all by itself, it was as obvious a lie as the one about the synagogue put on fire by the Jews. But within reach of the blind spot indicating father's disappearance, there was no adequate language. It was taken for granted that one couldn't have said: Father has been taken away in the middle of the night without having committed anything. Only the closest of friends could share this truth. With all the others one continued to weave the garment of lies called «gone on vacation». Even the genuinely ignorant knew then, fell silent, left, and did not come back. Nobody cared to know more about this departure. Was it prohibited to speak about it? It was not written on the wall. But since the deportations never had happened officially, they simply never occurred. Theses events never had become part of everyday life in the city and consequently *could not* have taken place. But to insist on something that could not have taken place would have comprised a dangerous challenge. It would almost have been a lie, a slander threatening the security of the state.

Even though nothingness seemed to have swallowed father, it spew him out again after a couple of weeks. Early one morning, still in the dark, he came to the house, rang the bell, climbed up the stairs, and stood in the kitchen. I saw father again. He had taken off his coat and stood there. Someone stood there. He opened his collar because he was uneasy. He took off the jacket because he wanted to do something. He looked us into the face and we looked him into the face. To us, his face showed bones with skin on top. The teeth had become longer. The eyes stood close to one another, frightful and full with obsequiousness. Thrashing had been written into the face. And above the beaten gaze the scull was bald.

The man had been shorn. He looked into our faces and apologized. He apologized because he had changed, because he stood unreasonably pitiful right in front of us in the middle of the kitchen. Beaten. Deplorable. And without hair. Humiliatingly shorn and pitiful. A wretched figure claiming to be a father, a shorn father. Something like this could now be.

( ) I embraced the thin neck and kissed father's cheekbones while he started to weep. But I laughed, we all laughed around the event, surrounding it with the joy of reunion. Father washed himself, had a shave, and drank coffee. The best of friends came, faltered, gulped, acted a part, and drowned the sight with joy.

The next couple of weeks without hair were difficult to get through. Father couldn't take off his hat, not beyond the closest circle of friends. After all, somebody walking around shorn was unheard of in Germany. That only happened to prisoners, but never to respectable citizens. Walking around like that, in a manner not possible in Germany, was itself an accusation, implied resistance, and was life-threatening. Nobody came back from a vacation with a shorn head.

But bristly hair grew again and grew longer. Eventually, it would lay on the side, having turned white, and covering the head. The beating and other issues kept silent, while written all over his face, were finally covered up. But some bald spots remained forever.

Now father knew that beating the drum for Germany in the trenches of the Great War had been in vain. The path of the ancestors, away from the village and into the city, revealed itself as a blind alley and came to an end. The shadow of the nearby Christian place of worship did not protect his house. The municipality of Freiburg declared his years of public years an error. Returning from the concentration camp to his art nouveau furniture did not symbolize a return home; here, only death was waiting. Father had to leave his house, his town, his country, without even having time to direct thoughts and feelings towards this unthinkable event. There was nothing to mourn for and no tragedy became part of an uncertain future. Everything happened rationally and was entirely simple. Father and mother had to flee. They escaped during the last night of peace on the last train from Germany. They carried ten marks in their pockets and a small suitcase each, unaware, *still* unaware of the luck granted to them. The God of the Jews once more presented them with the possibility of flight. They were not struck dead, gassed, and burnt: they could flee.

They hurried through already dark streets to the station. Racing with the radio announcers behind illuminated windows, racing with the war. And the last train at night still going on schedule was bouncing them out, over into freedom. (Paepcke, 1989: 55-9, my translation).

Max Mayer, who we read about in this later narrative by his daughter Lotte, was a merchant dealing with leather products and a speaker of the oppositional Social Democrats in the local city council through the Weimar years (Cf. also Böhme and Haumann, 1989).

12

Reading National Socialism as a «pathomorphology of modernity» (Peukert, 1987: 15, translation modified; also Paul, 1990: 13) in the archeological manner presented here makes understanding more difficult and identification at times painfully impossible, but harbours the possibility of truly unexpected insights. And we should *use* these insights creatively. Thus, rather than remaining trapped in a new frame, we should *deconstruct* incessantly. Is it a mere coincidence, we might ask, that our conventional frame around National Socialism so effectively left unstained that part of modernity which was necessary for a continuation of capitalism (in both its private and state-operated mutations) after World War II? All the values, all the gadgets we commonly associate with the economic boom of the 1950's, namely «home, car and trailer, radio and TV, cameras, kitchen appliances, detergents, hygiene, cosmetics, etc.» (Schäfer, 1981: 117, my translation) were *there* already, ready to be used for the stabilization of evil. Small wonder, then, that even the rhetoric survived: the concept of a «German economic *miracle*», so prominent during the '50's, «was coined already during the second half of the 1930's» (Dreßen, 1986: 262). And as «miracles» happen, they become a rather useful form of legitimation. Mass participation or mass consumption; but that, as the saying goes, is *another story* indeed. Another frame, or is it not?

Acknowledgements: This paper was first presented as a talk at the annual meeting of American Geographers in San Diego, April 1992. I would like to thank all the remarkable people in Freiburg who made working there truly enjoyable: E. Klaiber, Dr. H. Schadek, the late W. Vetter, P. Kalchthaler, F. Göpferich, and P. Bert. Also, I would like to thank D. Holdsworth for teaching me the joy of accepting certain risks that come with empirical research.

Notes :

1 This train of thought obviously owes its share of debts to both Ed Soja's *Postmodern Geographies* and Henri Lefebvre's *The Production of Space* within the social sciences, as well as to Jacques Derrida's *Glas* and Jean-Luc Nancy's *The Inoperative Community* within philosophy. Neither of them, of course, is to blame for my rather deliberate use of their work.

2 This line of argumentation, which emphasizes the *longue durée* of industrial modernity as the major undercurrent throughout the Nazi years, was first developed by Dahrendorf (1965), Parsons (1968), and Schoenbaum (1968). Marginalized since then, it erupted with sudden force during the so called *Historikerstreit* within German academic history in the mid- to late 1980's, producing a host of different analyses of *Nazism* or partial aspects thereof (Cf. Peukert, 1982, Reichel, 1991, Westphal, 1989, and Zitelmann, 1991, to name but the most visible of recent publications). Even though only few would be be willing to assert with Michael Prinz that «National Socialism was based on modernity and [that] there are no convincing reasons to accept that it ever wanted to liberate itself from that basis», 1991: 323, my translation), there is clearly a disenchantment with the inexplicable «singularity» of National Socialism being expressed today, in Germany and elsewhere. And yet, once more content with a «general» explanation of history, most of the writing arising from this disenchantment lends its voice all too easily to an everpresent conservative desire to relativize German history altogether. As a *possible* answer to the problems born of «modernity,» National Socialism is believed to be contextualized enough to be relieved of having to bear responsibility for any of the artrocities commited in its name. *On the contrary:* contextualizing Nazism, as I hope to demonstrate here, renders the possiblity of «totalitarian» «answers» omnipresent, thus cautioning us against any secured position from which to judge. There is, simply speaking, no inherent guarantee against fascism, no matter what society we are analyzing.

3 Even though the aeroplane was by far the most powerful image of modernity in those days. Explicitly *connecting* spaces, the aeroplane served its function of a «neutral» means in the silent service of a unifying fascination in a host of different contexts as well. If these «examples» were hitherto confined to contexts *not* mentioned here, this only stresses the importance of the proposed de- and re-contextualization attempted here.

Primary Sources

Freiburger Zeitung
Breisgauer Zeitung
Freiburger Tagespost
Volkswacht
Der Alemanne
Freiburger Theaterblätter
Vereinsblatt der Freiburger Turnerschaft
Stadtarchiv Freiburg im Breisgau

Universitätsarchiv Freiburg im Breisgau
Badisches Staatsarchiv Karlsruhe
Archiv der Firma Rombach&Co.
Archiv Vetter

References

Abraham, D., 1989, 'State and Classes in Weimar Germany', *Radical Perspectives on the Rise of Fascism in Germany, 1919 - 1945* 1989, ed. M.Dobkowski, I.Wallimann, New York, N.Y. : Monthly Review Press

Adorno, T., 1951, *Minima Moralia. Reflektionen aus dem beschädigten Leben* , Frankfurt an Main : Suhrkamp (translated into English as *Minima Moralia. Reflections from Damaged Life* trans. E.F. Jephcott, London : Verso, 1987)

Arendt, H., 1963, *Eichmann in Jerusalem. A Report on the Banality of Evil,* New York, N.Y. : Viking Penguin

Arnoldt, F., 1985, *Anschläge. 220 Politische Plakate als Dokumente der deutschen Geschichte* München : Langewiesche-Bradt

Benjamin, W., 1969, 'Theses on the Philosophy of History', *Illuminations* trans. H.Zohn, New York, N.Y. : Schocken

Blanchot, M., 1987, 'Everyday Speech', *Yale French Studies, 73*

Bloch, E., 1962, *Erbschaft dieser Zeit,* Frankfurt am Main : Suhrkamp (translated into English as *Heritage of Our Time* trans. N. and S., Plaice, Berkeley : University of California Press, 1991)

Böhme, R., and Haumann, H., 1989, *Das Schicksal der Freiburger Juden am Beispiel des Kaufmannes Max Mayer,* Freiburg i.Br. : Schillinger (Stadt und Geschichte. Neue Reihe des Stadtarchivs Freiburg im Breisgau, Heft 13)

Bräunche, E., 1983, 'Die Reichskristallnacht in historischer und kriminologischer Sicht', *Zeitschrift des Breisgau-Geschichtsvereins Schau-ins-Land* 104

Bräunche, E. *et.al,.* 1983, *1933. Machtergreifung in Freiburg und Südbaden,* Freiburg i.Br. : Schillinger (Stadt und Geschichte. Neue Reihe des Stadtarchivs Freiburg im Breisgau, Heft 4)

Broszat, M., 1981, *The Hitler State. The Foundation and Development of the Internal Structure of the Third Reich* trans. J.W.Hiden, London and New York : Longman

Buck-Morss, S., 1989, *The Dialectics of Seeing,* Cambridge, MA : MIT Press

Canetti,E., 1980, *Die Fackel im Ohr. Lebensgeschichte 1921-31,* München : C.H.Hanser (translated into English as *The Torch in my Ear* trans. J.Neugroschel. New York, N.Y. : Farrar Straus Giroux, 1982)

Dahrendorf, R., 1965, *Gesellschaft und Demokratie in Deutschland,* München, : C.H.Beck. (translated into English as *Society and Democracy in Germany* Garden City, N.Y. : Doubleday, 1967)

Dear, M., 1988, 'The Post-Modern Challenge. Reconstruction Human Geography', *Transactions. Institute of British Geographers* 13

Deutschland-Berichte der Sozialdemokratischen Partei Deutschlands SoPaDe 1980 ed. K.Behnken, Salzhausen, Frankfurt am Main : Zweitausendeins

Dreßen, W., 1986, 'Modernität und innerer Feind' in Boberg, J., et. al., eds., *Die Metropole,* München : C.H. Beck

Eichberg, H., 1989, 'Lebenswelt und Alltagswissen', *Handbuch der Deutschen Bildungsgeschichte,* C.Berg *et. al.* eds., Volume 5, München : C.H.Beck

Eksteins, M., 1989, *Rites of Spring,* Boston : Houghton Mifflin

Gärtner, K. ed., 1934, *Heimatatlas der Südwestmark Baden,* Karlsruhe : Künstlerbund

Gay, P., 1968, *Weimar Culture. The Outsider as Insider,* New York, N.Y. : Harper&Row

Grill, J.H., 1975, *The Nazi Party in Baden 1920-45,* Chapel Hill, NC : U of North Carolina P

Grünberger, R., 1971, *A Social History of the Third Reich*, London : Weidenfeld&Nicholson
Hamelmann, B., 1989, *Helau und Heil Hitler. Alltagsgeschichte des Fasnacht 1919-1939*, Eggingen : Edition Isele
Henning, E., 1975, 'Faschistische Öffentlichkeit und Faschismustheorien', *Ästhetik und Kommunikation* 6, 20
Herf, J., 1984, *Reactionary Modernism. Technology, Culture, and Politics in Weimar and the Third Reich*, Cambridge : CUP
Kern, S., 1983, *The Culture of Space and Time*, Cambridge, MA : Harvard University Press
Kracauer, S., 1971, *Die Angestellten*, Frankfurt am Main : Suhrkamp
Lacoue-Labarthe, P., 1989, 'Neither an accident nor a mistake', *Critical Inquiry* 15, 2
Lacoue-Labarthe, P. and Nancy, J.-L., 1990, 'The Nazi Myth',*Critical Inquiry* 16, 2
Lane, B.M., 1985, *Architecture and Politics in Germany, 1918 - 1945*, Cambridge, MA : Harvard UP
Lefebvre, H,. 1991, *The Production of Space*, Cambridge : Blackwell
Mittendorff, W., 1979, 'Als die Synagogen im Breisgau brannten...', *Freiburger Almanach* 30
Mittendorff, W., 1988, 'Die Reichskristallnacht in Freiburg', *Zeitschrift des Breisgau-Geschichtsvereins Schau-ins-Land* 107
Musil, R., 1965, *Der Mann Ohne Eigenschaften*, Hamburg : Rowohlt (translated into English as *The Man Without Qualities* trans. E.Wilkins, E.Kaiser London : Secker&Warburg, 1965)
Nipperdey, T., 1980, '1933 und die Kontinuität der deutschen Geschicht', *Die Weimarer Republik* ed. M.Stürmer, Königstein : A.Hain
Paepcke, L., 1989, *Ein kleiner Händler der mein Vater war*, Moos und Baden-Baden : Elster
Parsons, T., 1968, 'Demokratie und Sozialstruktur in Deutschland vor der Zeit des Nationalsozialismus', *Beiträge zur sozialen Theorie*, Neuwied : Luchterhand
Paul, G., 1990, *Der Aufstand der Bilder. Die NS-Propaganda vor 1933*, Bonn : J.H.W. Dietz Nachfolger
Peukert, D., 1982, *Volksgenossen und Gemeinschaftsfremde. Anpassung, Ausmerze und Aufbegehren unter dem Nazionalsozialismus*, Köln : Bund. (translated into English as *Inside Nazi Germany. Conformity, Opposition, and Racism in Everyday Life*, trans. R.Devenson. New Haven, CN : Yale UP, 1987)
Peukert, D.and Reulecke, J., 1981, *Die Reihen fest geschlossen. Beiträge zur Geschichte des Alltags unterm Nationalsozialismus*, Wuppertal
Quilitzsch, K., 1968, *Die Antisemitische Politik des Nationalsozialismus im Spiegel der Freiburger Tageszeitung Der Alemanne*, unpublished ms, Universität Freiburg im Breisgau
Prinz, M., 1991,, 'Die soziale Funktion moderner Elemente in der Gesellschaftspolitik des Nationalsozialismus', *Nationalsozialismus und Modernisierung* ed. M.Prinz, R.Zitelmann, Darmstadt : Wissenschaftliche Buchgesellschaft
Reichel, P., 1991, *Der Schöne Schein des Dritten Reiches. Faszination und Gewalt des Faschismus*, München : C.Hanser
Salb, T., 1991, *Das Stadttheater Freiburg in der Zeit des Nationalsozialismus*, unpublished Diss, Universität Freiburg im Breisgau
Schäfer, H.D., 1981, *Das Gespaltene Bewußtsein. Deutsche Kultur und Lebenswirklichkeit 1933-1939*, München : C.H.Hanser
Schnabel, T., 1983, 'Von der Splittergruppe zur Staatspartei. Voraussetzungen und Bedingungen des Nationalsozialistischen Aufstiegs in Freiburg im Breisgau', *Zeitschrift des Breisgau-Geschichtsvereins Schau-ins-Land* 102
Schnabel, T., 1986, 'Freiburger Pressekampf zu Beginn des Dritten Reiches, Teil 1', *Freiburger Almanach* 37
Schnabel, T., 1987, 'Freiburger Pressekampf zu Beginn des Dritten Reiches, Teil 2', *Freiburger Almanach* 38
Schoenbaum, D., 1966, *Hitler's Social Revolution. Class and Status in Nazi Germany*, Garden City, N.Y. : Doubleday
Schor, N., 1992, '*Cartes Postales*, Representing Paris 1990', *Critical Inquiry* 18, 2

Soja, E. 1989 *Postmodern Geographies,* London, New York, N.Y. : Verso
Stern, F., 1984, 'Der Nationalsozialismus als Versuchung', in Stern, F., Jonas, H., eds.,
    Reflexionen finsterer Zeit, Tübingen : J.C.B. Mohr
Strohmayer, U. and Hannah, M., 1992, 'Domesticating Postmodernism', *Antipode* 24, 1
Theweleit, K., 1980, *Männerphantasien,* 2 Volumes, Reinbek bei Hamburg : Rowohlt
    (translated into English as *Male Phantasies* trans. S.Conway, Minneapolis : University
    of Minnesota Press, 1987)
Westphal, W., 1989, *Werbung im Dritten Reich,* Berlin : Transit
Zitelmann, R., 1991, 'Die Totalitäre Seite der Moderne', *Nationalsozialismus und
    Modernisierung* eds., M. Prinz, R. Zitelmann, Darmstadt : Wissenschaftliche
    Buchgesellschaft

Ulf Strohmayer
University of Wales, Department of Geography
Lampeter, Dyfed SA48 7ED
United Kingdom

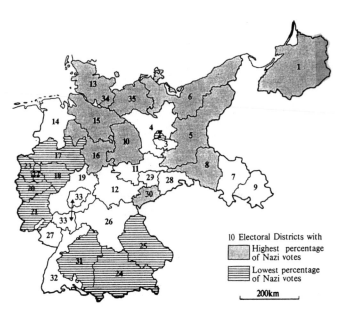

Figure 10.1: Election Results July 1932 (*«Länder»*)
source: Broszat (1981, xvi).
note: Baden is *Land* #32 on the map above

Figure 10.2: Local Advertisement
source: « Breisgauer Zeitung», 6 February 1933, 5.
translation::«Zeppelin is coming!»

Figure 10.3: Local Advertisement
source: «Breisgauer Zeitung» 9 February 1933, 5
translation: «Watch out tomorrow! Zeppelin is coming»!

Figure 10.4: Local Advertisement

source: «Breisgauer Zeitung», 10 February 1933, 5

translation:: «Zeppelin Journeys around the World, 264 real photos in every pack of CLUB, Germany's truly best 3.5 Pfenning cigarette»

Figure 10.5: Local Newsreport

source: «Freiburger Zeitung», Bilderschau, 29 September 1928, title

This picture captures the one and only occasion at which a *Zeppelin* was actually sighted *over* Freiburg. On its way from Friedrichshafen to the United States where it was to be handed over to the US government as part of the reparations fixed in the treaty of Versailles, this vast structure became more than just «the talk of the town», rather, it epitomized for the whole region the possibility of an encounter with «modernity», no matter how fleeing. The medium itself is noteworthy: beginning in late March 1924, the *Bildschau* was the first weekend edition «journal» in Germany to be distributed together with the Sunday edition of the *Freiburger Zeitung* (which was then one newspaper among four). The development of leisure time after the hyper-inflation of 1923 once again goes hand in hand with commercialization.

Figure 10.6: Goebbels arriving at Freiburg Airport, 16 June 1934
source: Municipal Archives Freiburg («Stadtarchiv Freiburg»), M 7092-20

Figure 10.7: Papal Nuncios Pacelli arriving at Freiburg airstrip, probably June 1929
source: Municipal Archives Freiburg («Stadtarchiv Freiburg»), M 2-127a-19

Figure 10.8: Official Poster of Freiburg Airport, 1933
source: Municipal Archives Freiburg («Stadtarchiv Freiburg»), M 7092-108
translation: «Regular Airtrafic with German Lufthansa to all continental airports. Fastest connection from Freiburg to Stuttgart -München-etc.»

Figure10.9: Advertisement, German Lufthansa, 1936
source: W.Westphal (1989), 58.
translation: «Visit the happy Germany»

Figure 10.10: Political Advertisement
source: «Der Allemanne», 27 July 1932, 9
translation: «3 Gigantic Election Rallies in Baden. Adolf Hitler speaks...»

Figure 10.11: Public Poster, 1938
source: Theweleit (1980, Vol.2, 88)

Figure 10.12: Official Advertisement 1940
source: Official Addressbook, Freiburg im Breisgau
(Municipal Archives Freiburg)
translation: «Honorary Citizen of Freiburg ...»

annes — war Vermittler aller Dollaran-
ihen und hat vermutlich erhebliche Provi-
onen dafür eingesteckt. Erst vor kurzer Zeit
urde dieser schwarze Herr von seinem
reund, Dr. Schlegel, wieder nach Newyork
eschickt, um dort Verhandlungen zu führen.
ahlreiche angesehene Persönlichkeiten haben

Alles wählt Hitler

ich wiederholt bemüht den Landeshauptmann
von seinem Freund Dr. Edlinger abzubrin-
gen. Es war vergebens. Es wurde nicht
nur nicht fallen gelassen, sondern auch noch
sein **Bruder bei der Bundesländer-Versiche-**
rung in Linz angestellt.

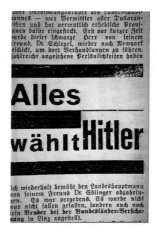

Figure 10.13: Political Advertisement, 1932
source: «*Der Alemanne*», 3 April 1932, 4
translation: «Everyone votes for Hitler»

Wer ist der beste

Elektriker der Welt?

Adolf Hitler!

Denn er hat:

Deutschland gleichgeschaltet
Polen ausgeschaltet
Rußland umgeschaltet
Die Tschechei eingeschaltet
Österreich angeschlossen
Röhm und Genossen geerdet
Die Juden isoliert
Spanien mit Starkstrom versorgt
England elektrisiert
Die Welt in Hochspannung versetzt
Dänemark und Norwegen neue Sicherungen eingesetzt
Holland, Belgien und Frankreich den Strom abgestellt
Die Schweiz mit Wechselstrom versehen
Und dabei **niemals Kurzschluß gehabt**!

Figure 10.14: Political Leaflet, 1940
source: Municipal Archives Freiburg («Stadtarchiv Freiburg»),M 2-32 (Nachlaß Winter)

This leaflet, which was distributed and posted in Freiburg after the «victorious» *Blitzkrieg* against France, is probably the boldest and most cynical «link» between modernity and National Socialism. Playing on various metaphorical employments of electricity, the leaflet states: Who is the best electrician in the world? Adolf Hitler! Because he has: brought Germany into line; switched off Poland; converted Russia; switched on the Czechs; connected Austria; grounded Röhm and his comrades; isolated the Jews; supplied Spain with heavy current; electrified England; brought high tension into the world; installed new fuses in Denmark and Norway; cut off the current in Holland, Belgium and France; brought alternating current to Switzerland; and never was short-circuited once!»

Figure 10.15: Commercial Propaganda, 1936
source: «Freiburger Zeitung», 17 November 1936, 11

Figure 10.16: Commercial Propaganda,
1936
source: Westphal (1989, 59)

Figure 10.17: Political Propaganda, 1936
source: Arnoldt (1985), 143
translation: «All of Germany listens to the *Führer* on
the utility radio set»

Figure 10.18: Touristic Propaganda, Baden, 1934
source: Schäfer (1981, picture 21)
translation: «3 weeklong trip of the *Badener* to Amerika», «As inexpensive as never before due to the weak dollar!»

Figure 10.19: Industrial Propaganda, 1927
source: «*Volkswacht*» 20 April 1927
translation: «Schwarzwald Industrial Fair»,«Man is lazy by nature - thus follow the cuckoo clock's example, which, never tired, always punctual, never takes a break»

Figure 10.20: «Freiburger Zeitung» 14 November 1938
translation: «England's new gigantic civil aeroplanes - the fastest connection between London and Paris will soon be employed for travel services. The latest De-Havilland aeroplane has been finished in London which covered the distance between London and Paris in 64 minutes. It is a 15-ton-aeroplane capable of carrying 22 passengers. The four engines develop 2400 hp.»

«No more Jews at German Universities.»

Figure 10.21: Freiburg Synagogue, 10 November 1938
source: Municipal Archives Freiburg («Stadtarchiv Freiburg»), M 7092-103

PART IV

ECONOMICS

11. Contemporary acceleration : world-time and world-space

Milton Santos

1 Contemporary acceleration

Accelerations are culminating moments in history which seem to concentrate powers that explode to create something new. The run of time, as mentioned by Michelet (preface to his History of the XIXth century) is marked by these great commotions which at times can seem meaningless. Thus, each time that we believe to have reached a definite stage, the unusual, the strange strikes us. The problem is to understand and find a new set of concepts able to express the new order created by this «acceleration».

Our contemporary acceleration is not at all different. It tends to be an object of metaphoric constructions for, to quote Jacques Attali, we fully live in the era of signs, after having lived in the era of the gods, of the body and the machine. Symbols confuse us because they have taken the place of true things.

The first temptation when faced with acceleration is to adore -faintly or strongly - the underlying velocity. In fact, speed was precisely what struck those who first witnessed the appearance of railways, of steam boats, and who thus, at the end of the XIX century and the beginning of the XIX, were reached by the diffusion of automobile, airplane, of wireless telegraph and submarine cable, of telephone and radio.

But why limit acceleration to velocity «stricto sensu» ? Contemporary acceleration has imposed new rhythms (on)to the movement of bodies and the transport of ideas, but it has also added new ingredients to history : a new evolution of potencies and productivity, the use of materials and new forms of energy, the mastering of the electro-magnetic spectrum, demographic expansion (the world population triplicates between 1650 and 1900, and again triplicates between 1900 and 1984), urban explosion, consumption explosion, exponential multiplication of the number of objects and words. And behind all of these

G. B. Benko and U. Strohmayer (eds.), Geography, History and Social Sciences, 171–176.
© 1995 *Kluwer Academic Publishers. Printed in the Netherlands.*

accelerations is the evolution of knowledge, this wonder of our times. Contemporary acceleration is for that very reason also a result of inventions, of banalizations and of premature aging of engines with their hallucinating succession. Nowadays, we find superposition, concomitance of accelerations. Thus the feeling of a fleeting present. This feeling of ephemera is not an exclusive creation of velocity. It comes also from the empire of image, and the way communication, which today belongs almost exclusively to the medias, deliberately impeaches the notion of duration and the logic of sequence.

These times of paradoxes alter the perception of history and confuse the spirits, and thus make place for metaphors to dominate the present discourse on Time and Space The conceptual challenge has to be faced now. Contemporary acceleration must be seen as a coherent moment of history. To understand it, it is necessary and urgent mentally to reconstruct the elements that compose our time and make it different from preceding times.

2 World time, world space

Can we imagine a world time whose Other would be a world space ? A world space resulting from the unfolding of world time ? This would suppose that world time really exists, as well as the world. However, we know that the world is but for others, not for itself since it only exists as a latency.

There is today a world clock, result of technical progress, but world time is an abstraction, except as a relation. And while we certainly have something like a universal time, it remains a despotic time, an hegemonic measure that controls the time of others. This despotic time is responsible for hierarchical temporalities, which conflict but converge miraculously. In that sense, all times are global, but there is hardly a unic world time. Space, too, is globalized, but is not a world space as a whole, except as a metaphor. All places are world places, but there is not a unic space. Actually, what is globalized are people *and* places.

What exist are thus hegemonic temporalities and non-hegemonic or hegemonized temporalities. The first of these are the action vector of hegemonic agents of economy, politics and culture, that is of society. Other social agents - hegemonized - are literally left with slower times. As to space, it adapts itself to the new era. In other words, it assimilates the components that make a given portion of territory the locus of productive and exchange activities for high and thus global-scale activities. Those places are hegemonic spaces taken by forces which regulate the action in other places.

3 Technosphere and psychosphere

Thus remade, space can be considered from the vantage point of both technosphere and psychosphere which together form the technico-scientific milieu. The technosphere is the result of the increasing artificalization of the environment. The natural sphere is being substituted for a technical sphere, a process that happens likewise in town and country. By contrast, the psychosphere is the result of beliefs, desires, wills and habits that inspire philosophical and practical behaviours, personal interrelations and communion with the Universe. The geographical milieu, the former «natural milieu» and today's «technical milieu», is nowadays tending to be a «technico-scientific milieu». This technico-scientific milieu is much more present as psychosphere than as technosphere.

Let us consider the case of Brazil. As technosphere, the technico-scientific milieu can be seen as a continuous phenomenon in most of South-East and South of the country, spreading other a large part of Mato Grosso do Sul. As psychosphere, it covers the whole country. Both facts have deep repercussions on economic practice as well as on social and political behaviours, and thus offer a new basis for the understanding of Brazil's process of regionalization. We can also propose to regionalize on the basis of rationality. Today, thanks to technical progress and contemporary acceleration, national spaces can be roughly divided into spaces of rationality on one side, and other spaces on the other side. Obviously, there are inumerous intermediate situation.

Throughout the centuries, human societies have been induced to a daily need for measure, standards, order and rationalization. The action sphere thus created has been studied by social scientists, except by geographers. Nowadays, this very space - the technico-scientific milieu - shows the same contents of rationality because its geographic objects carry an intentionality : their localisation, more than ever, responds to the design of social actors capable of rational action. This mathematization of space turns it favourable to the general mathematization of social life, according to hegemonic interests. In this manner the conditions for the maximum benefit for the strongest are established, which are also conditions for the most thorough alienation of all. Through space, the perverse side of globalization impoverishes and cripples.

4 Rationality, fluidity, competitiveness

Within these spaces of rationality, the market becomes dominating and tyrannical while the State becomes impotent. Everything is organized to let hegemonic flows run freely at the expense or against other flows. Consequently, this is also the reason why the State has to be weakened : to leave the way free for the actions of sovereign markets. No wonder, then, that the present period demands both fluidity and competitiveness. Help comes from outside societies through the strength of foreign theories and the power and the violence of money.

The need for fluidity rubs off boundaries, creates improved modes of transportation and communication, sets free the circulation of money (and later on the circulation of goods), and smoothens rugosities that might hinder the flux of hegemonic capital (for instance the «ejidos» in Mexico or «latifundios» in Brazil, both forms of huge land properties, are condemned by international financial institutions). Fluidity is the condition, but competitiveness is the basis of hegemonic action. This belief has its «Gospel and its evangelists»; this new bible is the WCI, i.e. the World Competitiveness Index produced by the World Economic Forum with the help of the Institute for Management Forum of Lausanne. This Index covers 34 countries on the basis of 130 criterions that measure enterprises' competitiveness within their respective environments (Petrella, 1991). International cities are furthermore classified according to their capacity for competition and their ability to attract activities which are of interest to the most aggressive businessmen.

Without the contemporary acceleration, the competitiveness which permeates the discourse and possible actions of governments and big enterprises alike would neither be possible, nor would it be without the recent technical progress and corresponding fluidity of space. In the present time, «competitiveness» takes the place of two former concepts : that of «Progress» at the beginning of the century, and that of «Development» after the 2nd world war. At the time, as Daniel Halevy reminded us, the debate was philosophical, even teleological. The notion of progress, for instance, carried the idea of moral progress as well. The debate around the notion of development and underdevelopment also had a clear moral connotation (see for instance Father Lebret's proposal for a human development).

In contrast the strive for competitiveness - as defended by governments, businessmen and executives from international institutions - does no longer need any ethical justification, nor does any other form of violence. Competitiveness is another word for war, now a planetary war, conducted by multinationals, foreign affairs bureaus international bureaucracy, and with the aid of some scholars from inside of outside the University. How can we wonder about war and violence, when the New World Order in construction is based on competitiveness without moral ?

5 Globalization and fragmentation

All attempts to build a unic world have led to conflicts because unification has been sought rather than union. The one side is characterized by a system of relations, which serves the greatest possible number of people and is based on realities and possibilities of a given historical moment. The other side represents a system of hierarchical relations, conceived to perpetuate a sub-system of domination upon other sub-system, for the benefit of a few ones. This latter situation is now prevailing : federative forces at a global scale are not operated by a will for freedom, but by a will for domination instead; not by a will for cooperation but by one for competition. This demands a rigid organization which interferes in every nook of human life. With such intentions, the forces of globalization lead to falsification, corruption, destabilization and destruction.

World dimension is the market. World dimensions are also the so-called international institutions, the supra-national institutions, world universities, demobilizing churches : the world as a big deceit factory. In such a world, the classical conflict between necessity and liberty is staged now between coercive organizations and the exercise of spontaneity. The result is fragmentation. whose most obvious dimension is the tribe - which unites people through resemblances - and the place - which unites people through cooperation within difference. The great revolt comes through space, through places where tribes discover they are not isolated. This place may be called Ngoro Kharabad as well as Los Angeles. The world of sick globalization is counteracted within a variety of different places.

All of this means that what globalizes, in fact divides; what is local allows union. A place is characterized as the extend of homogenous and solidary happening, whose constitution is the territorial configuration on one hand, and organization, regulation modes on the other hand. Place or region are no longer a result of organic solidarity, but of regulated or organized solidarity. No matter it be ephemer. Phenomenons are not defined exclusively through their duration, but also and especially through their structure. And finally : what is long, what is brief ?

Through a place we revisit the world and we adjust our interpretation, for in it predominates the deep, the permanent and the real over the movement, the transient, the foreign. Space appears to be a substratum that receives the new but resists to change, preserving the force of material and cultural inheritance, the forces of what spring from inside, quiet forces that await the occasion to express. Old cements become new cements : languages, religions, cultures, ways to contemplate Nature, the Universe, ways to look at oneself and look at the others. The basis of great transformations of the world map stems quite possibly from this kind of movement. From the need of an abstract State that is self reliant, we come to the necessity for a concrete State that would meet the deep wants of people. According to Edgar Morin (1965: 73), where we were supposed to have reached the «necessity of nation», this for him «still [remains] an obscure phenomenon».

But what does «Nation» mean today ? About old nations, we know something. They usually fit a piece of territory. About nations that are emerging in front of our eyes, what do we know ? Will they be in foreign lands, will they represent the re-disposition and reconstitution of old loyalties or inherited attributes ? Will the city be a nation ? Whatever it be, it seems that the basis for action-reaction is the daily shared space. Such questions also point towards the problem of scale for an efficient action on space. We may as well ask : where is the scale ? The spatial gap between action and its result is increasing. Thus scale can exist but has nothing to do with size (the old worries with distance) nor with organizational imposition. Scale is time.

6 The 5th dimension of space : the daily life

Space has a new dimension : density and depth of existence This new dimension is caused by the huge number and variety of objects, i.e. fixes which compose it, and because of exponential number of actions, i.e. flows crossing it. This implies a true 5th dimension of space. Time of shared daily life is a plural time, time within time. Today this is no longer exclusive of the city but concerns the countryside as well.

In analytical terms, spatialization means practical temporalization, which cannot exclude any actor. Existence is thus beaconed by the place, and in this sense at least it can be said that time is determined by space. Daily life is the 5th dimension of space and hence a task for geographers. In fact, time and space did not become empty or fancy as thought by Anthony Giddens, but, on the contrary and through daily life, time and space which contain the variety of things and actions, also contain infinite possibilities. Suffice it not to consider space as simple materiality, i.e. the domain of necessity but as the necessary theatre of action, i.e. the domain of liberty.

Life is thus not a product of technique but of Politics, it is the action that gives sense to materiality. Marcuse said (1970: 62) : «Today we are able to transform the world into hell and we are on the way to do it. But we are also able to do exactly the contrary». Human space has never been so important for the direction of history as it is today. If, to quote Sartre, «to understand means to change», to go ahead «further than myself», a refound geography inspired from present realities can be an efficient theoretical and practical tool to refound the Planet.

References

Attali, G., 1982, *Histoires du temps*, Paris: Fayard
Marcuse, H., 1970, *Five Conferences,* Boston: Beacon Press
Michelet, J., 1833, *Histoire de la France*, Paris
Morin, E., 1965, *L'introduction à la politique de l'homme,* Paris: Seuil
Petrella, R., 1991, 'L'évangile de la competitivité', *Le Monde Diplomatique* (Septembre), p. 32

Milton Santos
Rua Nazaré Paulista 163, apt. 84
05448-000 São Paulo
Brasilia

12. Structural change, theories of regulation and regional development

Mick Dunford and Diane Perrons

1 Introduction : Social and geographical change

The geography of any locality is an expression of natural and social processes of change. Any given territory both shapes the specific nature of social, economic and political change as it takes place and is itself shaped by such changes. There is, in other words, a dialectical relationship between society and space and between economic, social and political change and territorial development change.

In this chapter our intention is to set out some ideas as to how geographers can analyse these interrelationships. In the first part of the chapter we shall argue that at an abstract level there are two fundamental causal processes which shape the development of societies and their geographies: ideal and material causes, and that these processes stand in a dialectical relationship with one another. The interaction of these mechanisms is the source of processes of historical and geographical change. We argue however that it is essential to recognise several different temporalities and several different types of Time-Space differentiating unique sequences of events from cyclical movements and structural changes. In the second part we focus on cyclical changes and their association with transformations in the character of capitalism and the geographies of capitalist societies. There are two main approaches to long-run cyclical movements : neo-Schumpeterian approaches and theories of regulation. Our aim is to outline the central propositions of the latter. In the third part we consider the debates about the crisis of Fordism, neo-fordism and post-Fordism. The transformation in the map of regional economic development in Europe in the 1980s and early 1990s is, we argue, a result of an experimental search for new combinations of technological and social relationships. Differences in the success of different

G. B. Benko and U. Strohmayer (eds.), Geography, History and Social Sciences, 177–191.
© 1995 *Kluwer Academic Publishers. Printed in the Netherlands.*

experiments and of regional economies provide an indication of where the next phase of development will lead : in our view the direction is towards greater organisation rather than towards the competitive markets of the pre-Fordist era.

2 Structural, cyclical and territorial change

At a fundamental level a comprehension of the development of any society requires an understanding of how material needs and wants have been and continue to be produced and satisfied. Production and consumption are simultaneously physical and social activities. In the process of production people work directly with nature or more commonly nowadays with the products of past production and with tools, equipment and machinery to produce new items of value to themselves and to their societies. However while they are producing such goods and services they also enter into social relationships : in other words individuals and groups work with other people in a context defined by particular rights and obligations and in relationships of equality or hierarchy and of harmony or conflict. People also form part of an economic unit of production and reproduction either singly or as part of a household in which case the relations between the members of such units also need to be considered. Included in this case are the intra household division of labour and gender roles. Whatever their form these relationships are an important characteristic of the society in question. The state also plays a part in the determination of the nature of these relationships and in the process of production and reproduction by financing or underwriting economic and social activities.

The logic of social change is composed of two elements : ideal and material causality. Ideal causality refers to the way in which human beings are able to shape their own destinies. Human beings are able to conceptualise their future or elements of their future and either individually or collectively seek to bring their ideas into being. However human beings are not entirely free agents but subjects who must act in pre-given natural, technological and social conditions (Dunford and Perrons, 1983). Material causality refers to the effect of these natural and social conditions on human action and of natural and economic laws which shape the way nature and society can be modified. Since, finally, the external world which conditions human beings has increasingly been shaped and altered by human activity, human beings are increasingly conditioned not simply by natural conditions and processes but by what they have made of nature. In short human beings can shape their futures. They do so, however, in material and social circumstances not of their own choosing. And yet these circumstances which condition human action are in part a product of previous human activities.

In seeking to explain geographical developments or territorial change it is useful, therefore, to examine those theories that pay explicit attention to the structures within which human beings act but that also acknowledge the significance of the extent to which individuals have room to manoeuvre and of the capacities of human beings to shape those structures.

As societies evolve through time past systems of development are impressed on both the physical and social landscapes of any territory and newer forms of development evolve within these conditions to produce new systems of social, economic, political and territorial development.

Change takes place continuously, and much geographical and historical investigation is concerned with the details of these daily, monthly and annual events. However these daily events are overlaid by longer-term movements and by major changes of far greater magnitude and scope. Included are the structural changes in modes of production such as the changes from feudalism to capitalism. Within a mode of production there are also long-term cyclical movements such as long waves (Freeman, 1987) or regimes of accumulation (Aglietta, 1979). These structural and cyclical concepts are relevant not just for investigations of historical change but also for geographical studies. On the one hand there is a need to consider the structural and cyclical changes in Time-Space such as the rise and decline of world economies and of ideological divisions of space such as the 45-year East-West divide in Europe (Wallerstein, 1989). On the other hand modes of production such as feudalism or capitalism or regimes of accumulation such as Fordism or specific waves of growth are a result of and set the context for development within any territory.

To distinguish modes of production Balibar identifies three main criteria. These are (1) the nature of the connections between the direct producer and the means of production, (2) the form of surplus appropriation and (3) how production takes place - not what is made but how it is made. Feudalism, for example, was characterised by the direct producers, the serfs, (1) having access to the means of production within a given locality by right, (2) having the surplus appropriated from them very explicitly, by the legitimate use of force if necessary, in the form of rent in kind or later in a money form, and (3) having control over the labour process : in other words the direct producer controlled the pace and rhythm of work and they were not given instructions about how to work (Balibar, 1970).

In capitalism by contrast the direct producer, a free wage labourer, has no independent access to the means of production. The only way the workers are able to obtain what they need for survival is by working for a capitalist employer who owns and controls the means of production. The form of surplus expropriation is far from explicit. It takes place in the process of production by workers contributing much more to the value of the products than they receive in the form of wages. Moreover because the labour process is simultaneously the process through which the valorisation of capital occurs the capitalists or managers take on the role of determining how work should be carried out. (Indeed the study of work has become a science in its own right).

Yet while the capitalist mode of production is characterised by these three invariant relations of production, the specific form that these relations assume can and has varied through a succession of phases of capitalist development. In particular there has been a sequence of cycles : pre-industrial cycles of the merchant capitalist era with phases of expansion and crises of shortage characteristic of the agrarian societies of early modern Europe, and Kondratiev cycles of the industrial era. What is more the balance between capitalist and other social relations changes as capitalist societies evolve. (With, for example, the development of the welfare state there was a limited decommodification of labour-power and the growth of a set of economic activities that depended on collective property and non-market principles of allocation.) These cyclical changes within capitalist societies are a consequence of the fact that in capitalism the labour process is also the process of valorisation. Capitalism is therefore an economic system in which there are endogenous mechanisms that lead to almost constant innovation and change. As capitalists are in competition with other capitalists and in relations of conflict with their own workers, there is a constant need to change the way in which commodities are produced and to develop new commodities. Changes are introduced on a day to day basis. However there have also been changes of a more fundamental kind. These changes have been characterised in a number of ways : on the basis of the degree of competition, as with the transition from competitive to corporate and state monopoly capitalism, by the pre-dominant technological paradigm, as in long wave explanations, and by the existence of different regimes of accumulation. In geographical terms there have been parallel changes from small rural manufacturing workshops to large urban factories, suburban industrial spaces and new industrial spaces where in some cases small urban / suburban or even rural manufacturing workshops can be found.

Within these cycles there are a wide range of variations in social and territorial development, that reflect local circumstances. So far we have identified modes of production, long term cyclical movements and everyday particularities as elements of social and spatial change. To understand territorial change there is finally a need to study the many aspects of development which are often studied separately in different disciplines.

3 Explaining the cycles : Theories of regulation

Theories of regulation were developed to explain the apparent paradox between the inherent instability of capitalist society, as identified by Marx in his theories of the accumulation process, and the empirical reality in which relatively long periods of stable growth have occurred (Aglietta, 1979, Lipietz, 1987, Boyer, 1986). To explain this paradox a series of intermediate concepts were developed. These concepts lie between the theoretical concepts developed by Marx to analyse capitalist society and the special characteristics of capitalist society with its periods of stability and crisis. Included are the concepts of an industrial paradigm, a regime of accumulation, a mode of regulation, and a hegemonic structure.

The purpose of these concepts is to explain processes of socio-economic development that exhibit significant spatial and historical variations. What lies at the root of the movement from regular growth to stagnation and economic instabilities, why do growth and crises assume different intensities and characteristics in different nations and regions, and why does the character of crises differ from one historical epoch to another ?

The point of departure is an identification of fundamental social relations within which the strategies and actions of individuals and groups unfold. In capitalist societies two central social relations are identified. In the first place capitalism is a system of commodity production in which goods and services are produced for sale on the market : decisions about what to produce, how much, when and where are made not in accordance with a social plan but are the decisions of private and autonomous individuals, and the social validation of these decisions is determined *a posteriori* through the sale of the products concerned on the market. Incomes similarly depend on the market validation of the activities of the individuals who perform them.

What differentiates capitalist societies from other commodity producing societies is, however, the fact that the process of work itself is structured by the wage relation and the property relations on which it depends. In capitalist societies the commodity relation is generalised, and labour power itself also assumes the commodity form : as a result the process of labour in which use values are created is also a process in which the value advanced is increased and capital is valorised, while the self expansion of value or the production of capital emerges as the object of production. In theories of regulation the concern is with the concrete expression of these fundamental social relations. But in modern capitalist societies not all of the process of social reproduction is structured by these relations. Many goods and services are not commodities, not all work is performed by wage-earners, and some property is collectivised (see Dunford, 1988: 11-12). Over time these boundaries have shifted. As a result the changing articulation of capitalist social relations with other social relations is a fundamental factor in the analysis of concrete development processes (see, for example, Boyer, 1986: 42-5).

Regimes of accumulation

In capitalist societies the valorisation and accumulation of capital are the fundamental goals of economic activities, and so in theories of regulation the rate of profit and its determinants and the rate of investment are central variables. Viewed from the standpoint of an individual capitalist the success of any activity depends on the profit it yields. At the level of the economy as a whole, however, all production is interdependent, as are the activities of distribution, exchange, and consumption. An expansion of one industry depends on the expansion of industries supplying its capital goods and intermediate inputs, as well as on an expansion of demand for its output. The availability of inputs is conditional upon the supply decisions of other capitalists, while demand depends on the investment and production decisions of other firms, on the distribution of income in the form of wages, rent, interest and distributed profits and on the consumption decisions of individuals.

Accumulation and growth depend therefore on the existence of regular dynamic relationships among several elements : (1) the development of processes of production as processes of value creation, (2) the division of value added that shapes the reproduction of different classes and social groups, (3) the articulation of capitalist and non-capitalist sectors and corresponding value transfers, and (4) the composition of social demand on which the validation of production capacities depends.

The conditions necessary for equilibrium growth can of course be identified. Some of them are highlighted in Marx's reproduction schemas. In a capitalist society an equilibrated development of the different types of activity of which the social division of labour is composed is, however, not planned *a priori* or consciously organised. Only in the case of the detail division of labour within the enterprise does planning (of an authoritarian kind) prevail. The distribution of social labour among the various areas of economic activity is a result of the independent initiatives of thousands of capitalists. As the values in process represented by this wealth move through their circuits an interlacing occurs. On the one hand some goods and services are exchanged and the activities of making them are validated. On the other shortages and surpluses are identified. Adjustment depends on the market. Yet if adaptation depends on a sequence of *ex post* adjustments (which can themselves be destabilising), how do capitalist economies manage to grow in a balanced way ? At one level the acquired experience of a solution is itself one of the bases of a solution. The conditions inherited from the past and the expectations that earlier trends in the norms of production and consumption will continue are the foundations of a «social mould» which can sometimes be described as a regime of accumulation (see Lipietz, 1984: 5-6 and also Aglietta and Brender, 1984: 29-134).

A regime of accumulation is a systematic organisation of production, income distribution, exchange of the social product, and consumption. With the materialisation of a regime of accumulation economic development is relatively stable : changes in the amount of capital invested, its distribution between sectors and departments, and trends in productivity are co-ordinated with changes in the distribution of income and in the field of consumption.

Several schematic regimes of accumulation can be identified. In the 19th century a regime of extensive accumulation gave way to a regime involving a combination of extensive and intensive accumulation in which the investment of constant capital including investments in iron and steel, railway construction, and shipbuilding itself validated the growth of department 1. In the 1930s and after the Second World War in particular it gave way to a regime of intensive accumulation in which the conditions of existence of the wage-earning class were transformed through the articulation of mass production and mass consumption (see Aglietta, 1979: 66-72 and Lipietz, 1984: 6-7).

Industrial paradigms or trajectories

Of writers in the regulation tradition few use the notion of an industrial paradigm. The notion plays a much more important role in neo-Schumpeterian models, but it does converge with work on the wage relation and the process of labour (see, for example, Coriat, 1982, 1991).

The development of the process of labour has been punctuated by several major transformations. A phase of manufacturing was superceded by mechanisation. Mechanisation was given a new impetus by the development of scientific management or Taylorism whose goal was to accelerate «the completion of the mechanical cycle of movements on the job and (fill) the gaps in the working day» (Aglietta, 1979: 114-5), and in the 1920s the introduction of Ford's semi-automatic assembly line resulted in a mechanisation of transfer and a rationalisation of the flow of work. In the 1970s and 1980s with advances in electronics and in computer and communications technologies automation and systemofacture emerged as new principles of work organisation.

Successive waves of industrialisation were associated not just with changes in the organisation of work, in the skills and capabilities of workers and with the development of new machines but also with the development of new materials, a sequence of new products and a succession of leading sectors : after the Second World War, for example, growth was centred on the spectacular development of durable consumer goods and construction industries, while the growth of these sectors stimulated the demand for investment goods and for energy and intermediate goods such as steel and plastics (see Freeman, 1987 and Dunford, 1988). To explain the waves of accumulation and investment attention must be paid therefore to the development of industries, technologies and human skills that a concept of the forces of production denotes.

Modes of regulation

The concept of regulation is used to denote a specific local and historical collection of structural forms or institutional arrangements within which individual and collective behaviour unfolds and a particular configuration of market adjustments through which privately made decisions are co-ordinated and which give rise to elements of regularity in economic life (see Boyer: 1978, 28-9 and Boyer and Mistral, 1978: 2-5). Its role is twofold. On the one hand it expresses and serves to reproduce fundamental social relations. On the other it is through these structural forms that multiple, decentralised individual and collective rationalities with their limited horizons result in regular overall processes of economic reproduction. A mode of regulation defines therefore the rules of the game. In addition it allows a dynamic adaptation of production and social demand and, in capitalist societies, guides and stabilises the process of accumulation.

The institutional forms themselves are codifications of social relations, and the social relations that are relevant depend on the characterisation of the dominant mode of production. In the case of capitalism four major social relations are identified : (1) the monetary system and monetary mechanisms, (2) mechanisms connected with the regulation of the wage relation, (3) modes of competition within the capitalist sector and between it and other non-capitalist spheres, and (4) the character and role of the state.

In its widest sense the wage/labour relation, for example, includes the organisation of work, the determination of the length and intensity of the working day, the ways in which labour-power is recruited, the structure and acquisition of skills, conditions of employment, the factors that determine the level and distribution of direct and indirect wages, and the ways of life and modes of reproduction of the wage-earning class which are more or less dependent on the acquisition of commodities and the use of collective goods and services. Historical studies show that it has assumed very different forms in the course of capitalist development : a competitive regulation with very limited working class consumption of capitalist commodities, a Taylorist regulation in which work was transformed without a commensurate change in working class life styles, and a Fordist regulation in which new norms of consumption emerged in parallel with new norms of production (see Boyer, 1986: 49-50).

Theories of regulation differ fundamentally from neoclassical views of adjustment. In real economies, as opposed to the fictions of the neoclassicists, convergence on an equilibrium path is extremely improbable, while the adjustments that do occur and that do result in the reproduction of the overall system are quite different from the adjustments claimed to operate in competitive markets. What is offered, therefore, is an alternative to theories of individual choice and concepts of general economic equilibrium.

Underlying the concept of regulation is a distinction between social action and institutional arrangements or the social conditions of human activity. Conflicts between classes and social groups, the strategies of enterprises and financial institutions which themselves are rooted in the fractioning of capital into interdependent private units, and the relation of these conflicts and strategies with various markets and the state are expressed, in each historical situation, in a collection of institutional forms : (1) structural and legal constraints that require specific types of individual and collective action but do not determine them, (2) collective agreements, and (3) systems of values, shared expectations and rules of conduct, which have emerged out of a historical process of class conflict, inter-capitalist rivalry, and state action. Adjustments in each market accordingly depend on institutions or structures which possess a certain degree of autonomy and cannot be reduced to a global mechanism such as that of supply and demand.

The stability of regulation presupposes a certain inertia of structures and institutional arrangements. But stability is only relative. The process of regulation itself engenders permanent movements which continually modify the character of social relations, the intensity of conflicts, and the relations of strength. As a result a type of regulation can only be identified as a result of a process of schematisation carried out for the purposes of theoretical analysis. Actual processes not only differ in a multiplicity of ways but also are subject to a continuous dynamic caused by changes in the social and political relations underlying them.

At certain critical moments in history the process of development results in a bringing into question of previously formed constraints and opens up the question of new forms of overall reproduction. Out of this real dialectic operating in each social formation, and out of the conflicts between classes and political groups, the strategies of organised social movements, and political processes unfolding within the state itself, specific types of regulation were developed.

Hegemonic structures

In these moments of change in the direction of human social development there is not one but a whole range of different possibilities. Which ones succeed depends in part on the economic success of different models and in part on the strength of different strategic concepts, the influence of the social groups that support them, the construction of coalitions and the actions of the state.

Jessop (1983: 89-109, and 1990a and b) has suggested that state action is related to accumulation strategies on the one hand and hegemonic projects and associated alliance strategies on the other. An accumulation strategy is «a specific economic «growth model» complete with its various extra-economic preconditions and ..... the general strategy appropriate to its realisation».

A hegemonic project is a political, institutional and moral strategy which is economically conditioned and relevant but whose domain is civil society as a whole and not just the economic sphere. In it the general interest is identified with a general institutional and policy framework and with a general programme of action which on the one hand advances the long-term interests of the hegemonic class or class fraction and on the other enables some of the goals of other allied or potentially allied interest groups to be attained. Through a programme that has a material as well as an ideological content the construction and reproduction of wider social and electoral blocs is accordingly facilitated. In the years after the Second World War, for example, the establishment of a new model of development involved a class compromise centred on the ideological project of the Keynesian welfare state : the working class was incorporated into a state that advanced the interests of industrial capital with an offer of modernisation, social reform, individual consumption, equal opportunities, and steady economic advance (see Hirsch and Roth, 1986: 74-7).

Crises

Theories of regulation offer explanations not only of growth but also of crises and of the movement from one to the other. Several types of crises are identified (see especially Boyer, 1986: 60-72). Crises can first of all result from «external factors» such as natural disasters or wars (whether such factors are exogenous is a matter of some controversy).

In the second place crises occur as part of the normal development of a mode of regulation. In phases of faster growth a whole series of tensions build up which are resolved in phases of slower growth : an example is provided by Marx's account of the way in which investment slowdowns and increases in the size of an industrial reserve army enable employers to alter the distribution of value added in favour of surplus value and reimpose industrial discipline.

Structural crises on the other hand involve crises in a mode of regulation and crises whose roots lie in the exhaustion of a model of development. (The fact that crises are also associated with innovations and further transformations in the forces of production is usually given insufficient weight). Structural crises do not, in the view of most writers in this tradition, have regular causes. In 1929 a cumulative collapse was a result of the limits to self-centred accumulation in department 1 and obstacles to the growth of demand in department 2 (see however Duménil, Glick and Rangel, 1986 whose examination of data for the US indicates that the central cause was an insufficient rate of profit in the industrial sector), whereas the crisis of the 1970s was rooted in a fall in the rate of profit and the exhaustion of an industrial paradigm. What is more no automatic mechanisms ensure that structural crises will give way to new waves of expansion.

Theories of regulation have therefore evolved and involve not one but several approaches. In our view a synthetic approach that integrated technological and political factors into the initial system of concepts offers a means of strengthening and consolidating the initial advances. With these debates, however, the central ideas have also come under criticism. In these circumstances there is a clear need to identify very clearly the claims that these theories make and to delineate their domain identifying what is to be explained and what is not.

4 The crisis of Fordism and after

The crisis of Fordism

Working outside the regulation approach, although with a similar periodisation of economic change, Piore and Sabel (1984: 165-193) advance an influential account of the breakdown of mass production. In their view the demand for mass produced goods stagnated as markets in advanced countries were, it was argued, saturated, while consumers sought goods that were more diversified and had a higher design content. In this situation smaller and more flexible specialised enterprises that made more diversified goods and services and that employed skilled craft workers started to gain the upper hand and offered the prospect of a new model of development called «flexible specialisation». Moreover in place of vertical integration between plants producing different components for a particular product market transactions between independent firms were more likely. (If development does indeed proceed along this path, argued Piore and Sabel, the 1970s and 1980s will be seen in retrospect as a major turning point in the historical process of industrialisation : a second industrial divide in which industrial societies returned to nineteenth-century craft methods).

Theories of regulation, on the other hand, are associated with a different view that identifies two major factors. In the second half of the 1960s the rate of productivity growth declined and the capital-output ratio increased. Investment and capital intensification in department 2 slowed down, though the effects of this slowdown were offset in part by the acceleration in the development of the export sector. At the same time real wages increased and workers' struggles against Taylorist modes of work organisation grew in importance. Together these factors resulted in a decline in the rate of profit and a crisis of valorisation. Insofar therefore as there is a crisis of an industrial model its roots lie not in market conditions but in its capacities as a method of value production. (The change in the rate of profit varied from one sector to another. Investment was channelled in sectors whose growth potential seemed greater. As a result the situation of excess capacities and overinvestment was exacerbated in the wave of expansion that preceded the oil crisis).

At a later stage a second demand side was added and, in the subsequent evolution of the crisis, interacted with supply side factors. The causes of demand side difficulties were closely connected with processes of internationalisation : as wages increased and the rate of profit fell, investment at home declined, and the internationalisation of production accelerated. In these conditions there was an increase in unemployment. Job loss however did not at first result in a major drop in aggregate demand but in an increase in the costs of the welfare state that provided the unemployed with alternative incomes. It was with the rise of monetarism and the adoption of austerity measures that the situation changed and the second phase of the crisis was opened. Increasingly the export sector was the propulsive sector, and in 1977-80 wages in particular were seen as a determinant of competitiveness and not as an element of final demand for domestic firms. Internal demand stagnated further as, with the internationalisation of the recession, did world demand : in order to reduce its balance of payments deficit each nation sought larger wage reductions that its rivals, and to improve its capital account each nation introduced yet higher interest rates to attract international deposits. (The saturation and stagnation of markets and related economic instabilities that resulted were in turn at the root of the search for more flexible structures. What writers on flexible specialisation see as the cause of the crisis of mass production was therefore more a result).

As a result of the difficulties of accumulation the share of banking capital in aggregate surplus-value increased, as did other rentier and speculative incomes (see van der Pijl, 1984: xvii-xviii). Savings centralised in transnational United States banks were used to finance industrial development in peripheral and semi-peripheral countries, yet at the end of the 1970s the countries that did borrow found themselves squeezed between a stagnation of demand on the one hand and an increase in interest rates on the other (see Lipietz, 1985).

Subsequent stages in the unfolding of the crisis have also been traced. With respect to the concerns of this chapter what matters is that the account of the crisis of Fordism is very similar to classical Marxist accounts of crises of valorisation.

Out of crisis ?

One of the major current debates that has emerged out of theories of regulation involves arguments over the possibilities and relative merits of different ways out of the «double-sided» crisis of Fordism. In most cases what is envisaged is a further transformation of capitalism : just as in the face of earlier crises, the character and mode of operation of capitalism may, it is argued, be changed, adapting it to the causes of the current breakdown, extending its developmental potential, and altering the conditions of maturation of its contradictions. (The question as to whether the changes in social relations that will shape the world in the 21st century will entail further developments that move in the direction of a supercession of capitalist social relations is asked infrequently but is one that remains on the agenda).

To cope with the downturn in the rate of profit and the supply side difficulties added importance was given to automation as a new principle of work organisation. Automation is associated with two major changes in the sphere of production. On the one hand the machine system itself is more flexible. Instead of specialised machines and specialised workers, machines which can execute a variety of different operations and which can be switched from one to another quickly and under computer control are used. As a result the life cycle of the machine is delinked from the life cycle of the product, and production can itself be coordinated more closely with design and marketing. On the other it allows production to be controlled as it occurs : the allocation of work between work stations can be managed in real-time as can the management of inventories and the scheduling of production.

New principles of work organisation and in a more general sense new technologies do not impose new sets of social relations. A model of development is a question of politics and not simply a result of technological imperatives or the operation of the abstract laws of economics. The success of different sets of technological and social choices will however be reflected in unevenness in the character, speed and quality of development, while what happens in areas that are more dynamic will in the end have an important impact on ones whose growth is slower : uneven development is in other words one of the mechanisms through which the development of the forces of production is carried forward under capitalist social relations, and an index of the success (as measured in value terms) of different types of adjustment. (What is successful however need not be what is progressive).

What is clear at the moment is that there is no new hegemonic model of development. A question that does arise however is the following : with the new developments in the forces of production which societies will do better : societies that give enterprises and markets a free hand or societies that impose socially enforced social obligations on them ? Integrated companies, independent small firms or integrated networks of firms ? Societies whose enterprises use new technologies to deskill or enterprises that seek more consensual management styles ? Societies in short that are more market oriented or societies that are more organised ?

In an era in which neo-liberal perspectives are dominant there is strong evidence that countries that are more organised and where more offensive strategies of innovation in the fields of work organisation and worker involvement such as Japan, Germany, or Sweden have done better than more market-oriented economies such as the United States or the United Kingdom (see, for example Leborgne and Lipietz, 1988 and 1989).

What Leborgne and Lipietz point out is that responses to the supply-side problems at the root of the crisis of Fordism can involve one of two types of action : either an employer can attack the rigidity of the wage relation through attempts to increase the numerical flexibility of the workforce and the flexibility of wages, or an employer can choose to attack the problems caused by the Taylorist division of mental and manual work through attempts to involve a multiskilled and functionally flexible workforce in the quest for productivity and quality.

In the UK and the USA a neo-Taylorist approach seems to have been adopted with very few workers involved in the decision making of the firms or enterprises of which they are a part. Moreover relations between the firms themselves tend to be hierarchical with the larger firms using their market power over the smaller firms in their orbit. In Japan, on the other hand, workers in the more dominant enterprises, at least, seem to be actively consulted about the development of processes and activities within their firms and worker cooperation and involvement is seen to be an important element in ensuring that potential productivity gains within any new techniques are actually realised. The value placed on the role of these workers is reflected in higher wages and more secure working conditions. However not all the population is involved in firms of this kind so although there is involvement at the level of the firm for some workers, for the society as a whole there is a form of dualism often along gender and ethnic lines between those included and those excluded from the negotiated division of advantages found within the dominant firms' sector.

Towards the opposite end of the continuum starting from the neo Taylorist practices followed by the UK and USA and the compromises negotiated at a firm level in Japan there are the cases of Germany and Sweden where there is in the first case involvement at the level of the sector giving rise to more privileged conditions for workers belonging to particular sectors of the economy. The workers are involved in the decisions of their firms and between the firms there are some forms of cooperation or collaboration rather than pure market transactions as predicted in the flexible specialisation model, leading it would seem at the present to conditions more conducive to investment and faster economic growth and greater economic prosperity for the mass of the population. In the case of Sweden the collaboration between the firms is supplemented by forms of corporatism between the firms and state institutions or society. In these circumstances more stable conditions for the emergence of coherent spatial structures exist.

The point that is being made is that there are different ways out of the crisis. These choices are reflected in differences in the trajectories of regional development in advanced countries. What is more the competitive advantages of different choices is a major determinant of the map of economic development and of differences in output and income in different regions. So far those countries that have adopted more progressive strategies that have involved people in the decision making processes have been more successful in overcoming the crisis than those who have adopted neo-Taylorist models and where extreme social polarisation prevails (Benko and Dunford, 1991).

In the end, however, there is also a need to recognise that choices concerning the character of the wage relation, the organisation of inter-firm relations, mechanisms of income redistribution and of the role of the state can however not͡ be made in a vacuum in each nation. The outcome of the crisis will also depend, as theories of regulation indicate, on whether or not new international agreements come to rule over the laws of the jungle that prevail in international markets and on whether what is at present a predominantly zero-sum game can be transformed into a positive-sum situation.

References

Aglietta, M., 1979, *A theory of capitalist regulation*, London: New Left Books.

Aglietta, M. and Brender, A., 1984, *Les métamorphoses de la société salariale: la France en projet,* Paris: Calmann-Levy.

Balibar, E., 1970, 'The basic concepts of historical materialism' in Althusser, L. and Balibar, E., eds., *Reading Capital,* London: New Left Books,

Benko, G. and Dunford, M., eds., 1991, *Industrial change and regional development: the transformation of new industrial spaces,* London: Belhaven

Benko, G. and Lipietz A., eds., 1992, *Les régions qui gagnent,* Paris: PUF

Boyer, R., 1978, 'Les salaires en longue période', *Economie et Statistique,* 103 (September) 27-57

Boyer, R., 1986, *La théorie de la régulation: une analyse critique,* Paris: Editions La Découverte

Boyer, R., ed., 1988, *The search for labour market flexibility.The European economies in transition,* Oxford: Clarendon Press

Boyer, R. and Durant, J-P., 1993, *L'après-fordisme,* Paris: Syros

Coriat, B., 1982, *L'atelier et le chronomètre,* deuxième édition, Paris: Christian Bourgois.

Coriat, B., 1991, *Penser à l'envers,* Paris: C. Bourgois

Davis, M., 1986, *Prisoners of the American dream,* London: Verso

Duménil, G., Glick, M. and Rangel, J., 1987, 'Theories of the Great Depression: why did profitability matter?', *Review of Radical Political Economics,* 19, 2, 16-42

Dunford, M, 1988, *Capital, the state and regional development,* London: Pion.

Dunford, M. and Perrons, D., 1983, *The arena of capital,* London: Macmillan

Durant, J-P., ed., 1993, *Vers un nouveaux modèle productif,* Paris: Syros

Freeman, C., 1987, *Technology policy and economic performance: lessons from Japan,* Oxford: Frances Pinter

Freeman, C., 1988, 'The factory of the future: the productivity paradox, Japanese just-in-time and information technology' *PICT Policy Research Papers,* 3 (May) London: Economic and Social Research Council

Glyn, A., Hughes, A., Lipietz, A. and Singh, A., 1988, *The rise and fall of the Golden Age,* Cambridge: Department of Applied Economics Discussion Paper, 884

Hirsch, J., 1983, 'Fordist security state and new social movements', *Kapitalistate,* 10/11, 75-88.

Hirsch, J., 1984, 'Notes towards a reformulation of state theory', in Haenninen S.and Paldan, L., eds., *Rethinking Marx,* Berlin: Argument Verlag, 155-60

Hirsch, J. and Roth, R., 1987, *Das neue Gesicht des Kapitalismus.Vom Fordismus zum Post-Fordismus,* Hamburg: VSA Verlag

Jessop, B., 1983, 'Accumulation strategies, state forms, and hegemonic projects', *Kapitalistate,* 10, 89-111

Jessop, B., 1990a, 'Regulation Theories in Retrospect and Prospect', *Economy and Society,* 19, 2, 153-216

Jessop, B., 1990b, *State Theory: Putting Capitalist States in their Place,* Cambridge: Polity Press

Leborgne, D. and Lipietz, A., 1988, 'L'après-fordisme et son espace', *Les Temps Modernes*, 43, 501 (April) 75-114

Lipietz, A., 1979, *Crise et inflation, pourquoi?,* Paris: Maspero

Lipietz, A., 1985, *The enchanted world. Inflation, credit and the world crisis,* London: Verso

Lipietz, A., 1987, *Mirages et miracles. The crises of global Fordism,* London: Verso

Piore, M. and Sabel, C. F., 1984, *The second industrial divide: possibilities for prosperity,* New York: Basic Books

Storper, M. and Scott, A. J., eds., 1992, *Pathways to Industrialization and Regional Development,* London: Routledge

Van der Pijl, K., 1984, *The making of an Atlantic ruling class,* London: Verso

Wallerstein, I., 1988, 'The inventions of TimeSpace realities: towards an understanding of our historical systems', *Geography,* 289-297.

Mick F. Dunford
School of European Studies
University of Sussex, Arts Building
Falmer, Brighton BN1 9QN
United Kingdom

Diane Perrons
57, Upper Abbey Road
Brighton BN2 2AD
United Kingdom

13. Theory of regulation and territory : An historical view

Georges Benko

In the second half of the 1970s a new approach to economic problems developed in France, «the approach of regulation». During the 1980s it spread on the one hand from a little circle of French economists to other disciplines like geography and industrial relations, and on the other hand to other countries, in particular the Anglo-Saxon countries. The article which follows attempts to present the effects of this approach on those sciences which take account of the organisation of space : geography and urban, regional and international economics.

1 Origins and specificity of the theory of regulation

During the strong and regular growth in the 1960s, economic reflection turned essentially towards modelling, concerned with helping worried governments to avoid an overheating of economies while trying to maintain full employment. The economic difficulties of the early 1970s inspired a renewal of research interest in crises and saw notably the emergence of a wave of thought with multiple inspirations. The last decade marked a reinforcement of the disciplinary barriers, each of the social sciences seeking to reinforce its own foundations, beyond interdisciplinary declarations. On the contrary, the theories of regulation find their origin in the confrontation and the transformation of a series of tools and different approaches.

From Marxist theory they retain an interest in long movements, the idea of conflicts of interests which set groups of economic agents against one another, the historic vision of the means of production, but without holding on to the grandiose and erroneous dynamic that the successors of Marx piously admired.

G. B. Benko and U. Strohmayer (eds.), Geography, History and Social Sciences, 193–210.

The regulationist thought in essence proposes today a pattern sufficiently new and broad that one can not reduce it to neo-Marxism. It distances itself very notably from standard Marxism. From non-orthodox macroeconomics the influences came from Kaldor, Robinson and Kalecki (even more than from Keynes). The lessons drawn from this work are that strong and stable growth and full employment are the exception and not the rule. From the *Annales* School (particularly the work of Labrousse) the followers of regulation draw the following lesson : every society has a conjuncture et crises of its structure.

Law and political science have made their contribution to regulation theory by means of studies of institutions : «the rules and juridical institutions ... would not be a simple assembly of pre-existing economic relationships but allow them to be conceived and developed» (Lyon-Caen, Jeammaud, 1986). A second generation of work analyses with precision the genesis of the major crises and reflected on the logic of institutions and forms of organisation. These first elements will have to allow the study of the genesis, the expansion and the decline of institutional forms and the means of getting out of major crises. So far the programme has been retrospective and historic. It is also completed by a a more prospective second aim : to try to follow in real terms the reconstructions that herald a way out of the present crisis. It proposes to draw up a typology of institutional innovations, to clarify the determinants of the diffusion of new forms of organisation, to distinguish between local and global changes, and finally to define the structural compatibility of all the mutations which are happening. This renewed research on the mechanisms of regulation in the face of the extent of the mutations of which the crisis is the bearer opens up therefore a wider framework, that of institutional economics. The radical American analyses of the «Social Structure of Accumulation» followed a parallel path. Bowles, Gordon and Weisskopf analyse the crisis by means of a study of the general coherence of institutions, the social relationships of American capitalism and its progressive calling into question of its own success. They situate, like the followers of regulation, their analysis of the contemporary crisis in the wider framework of long cycles.

The concepts of «habitus» and «field» («champ» or «scope») are particularly well suited to the project of the school of regulation (Bourdieu, 1980: 113-120). According to the acceptance of the term in Bourdieu's sense «habitus» is a disposition of individuals, socially constructed, to «play the game» at the heart of the mode of regulation. «The permanent struggle at the interior of the field is the engine of the field. One sees in passing that there is no antinomy between structure and history and as such that which defines the structure of the field ... is also the principle of its dynamic. Those who fight for domination are ensuring that the field is changing, that it restructures itself constantly» (Bourdieu, 1980). This vision is found in the analysis of the slow transformation of the means of regulation, as far as the threshold, after which its structural stability collapses and that the question is raised about the redefinition of institutional forms, which themselves presume the recomposition of a myriad of fields and field areas and «habitus».

In the same line of non-orthodox methods, a wave of thought (open to the problems of cognition) wishes to develop a more precise economic representation than the conventional mechanism of co-ordination. That is «the economics of conventions» which prolongs the wave of regulation (at least across the relationship of frequent work) taking partial support from cognitive sciences. The economics of «conventions» is interested in the question of the convergence of anticipations based on the calculations of actors and the necessary existence of conventions which allow one to structure the changes. Among the principal representatives of non-orthodox thinking are Robert Salais, André Orléan, Olivier Favereau, and Laurent Thévenot, among others. The notion itself was introduced by D.K. Lewis in 1969. A general presentation of current research is in the work edited by A. Orléan (1994). The point of departure of this research is the identification of the obstacles to co-ordination which come from the non-fulfilment of a pure market logic : non-fulfilment of the competitive logic, of strategic rationality and of contracts. Before these limits it is necessary to elaborate a theory of collective action in context reflecting on the representation of the rules and models of the enterprise : the resources of co-ordination come from the collective cognitive devices, conventions and objects. One finds a spatial application of this line of reflection in the work of Salais and Storper (1993). We note that certain followers of regulation strongly criticise the theory of conventions and refuse all association with this wave of thought, such as Alain Lipietz (1994, 1995b). It is Lipietz who opens the way for regulationists towards a new horizon of reflection linking economy, politics, environment and ecology (Lipietz, 1992, 1993, 1995a).

This research is completed by other centres of reflection. Piore and Sabel (1984) influenced by their Parisian colleagues worked on industrialisation in the 1980s and introduce the term flexible accumulation; the Californian school of geography studied the spatial transformations associated with every big model of development and they highlighted the central question of the new productive systems likely to replace Fordism (Scott, Storper, Walker). Freeman, Perez, Clark and Soete, that is to say the group from the University of Sussex, bring forward particularly useful elements in considering that the long cycles correspond to a succession of «techno-economic paradigms» perceived as essential elements of the Schumpeter long wave theory. The evolutionists built up the notion of the national system of innovation in order to take into account the interweaving of technological, social and economic factors which are the basis of innovation (Nelson, 1993, Nelson and Winter, 1982). Dosi and his collaborators have a project to clarify the macroeconomic origin that define a means of development.

The list is long. Dunford seeks to interpret European integration, Jessop presents a critical reformulation of Fordism and post-Fordism. Mitchell, Commons and Veblen and the American institutional wave and the German school (Schmoller, Wagner) from the beginning of the century can enhance contemporary research. In Latin America an original branch of the theory of regulation is developing with leaders like Pinto, Ominami or Sunkel. Yet again in the framework of the association of a sociologist and an economist, Boltanski and Thévenot, (1987, 1991) open the way to an approach that no longer reduces the actors to agents dominated by exterior forces, but which studies them in the context of mastering their behaviour and their co-existence in the everyday world, this proposition integrates itself also in the vast project of the economy of

Figure 13.1 : The theory of regulation : sources and evolution (according to the idea of R. Boyer)

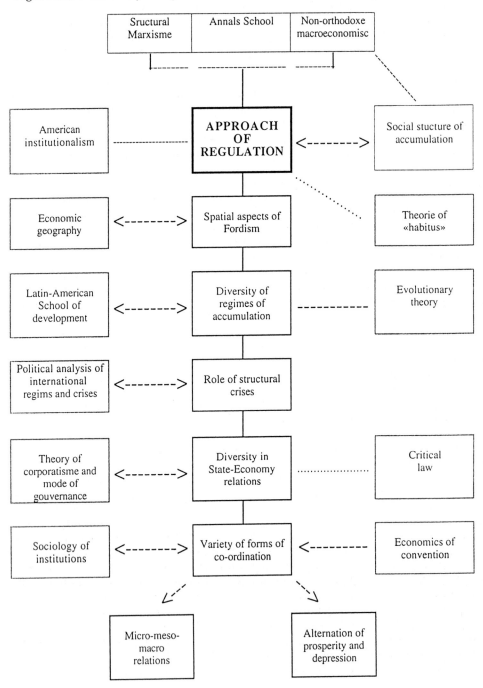

conventions. More recently the Japanese experience was confronted with the theory of regulation and in the 1990s a new animated debate appeared from the good «connoisseurs» of European regulationists like Yamada, Miyamachi, Itoh, Kato, Inoue and Yamada (Inoue and Yamada, 1995; Peck and Mijamachi, 1994).

Following this broad outline let us get back to the basic ideas. Initially it seems to me that there were some theoretical and political dissatisfactions by a group of researchers. The first dissatisfaction concerns the dominant theories of growth of the era, notably Solow, who tried to change a certain number of added factors to an exogenous postulated growth. Justly, one of the constituent contents of the School of Regulation consisted of integrating an endogenous «engine» into the expansion of capitalism capable of explaining at the same time the diversity of national trajectories and the reasons for which the figures of growth could have diverged for such long periods. The second dissatisfaction, fundamental to the regulationalist thinking, concerns the development of Marxism in the 1960s, the angle of the criticism of conventional theory, the affirmation of the deepening and inescapability of crises, of the lowering of the profit figures, of the monopolist capitalism of the State. These double criticisms vis-a-vis the neo-classicals and the orthodox Marxists gave the first momentum to regulationalists.

The approach of regulation is first of all an amendment of the structuralist approach. Instead of noting the permanence of the structures and evaluating their effect on the behaviour of agents, one questions the very stability of these structures. How are structures, in spite of their contradictory character, reproduced through the a priori divergent expectations, interests and actions of economic agents ? The answer is in the analysis of habits and institutional forms which induce or force the agents to behave in a way which is not antagonistic to the reproduction of the structure. This unity of «rules of the game» and procedure of resolution varies in time and space, in such a way that the structures can «function» in different ways, which are relatively stable between two crises. This method of functioning is called «the accumulation regime».

First it is necessary to say a few words about the origin of the word regulation. In France one of the first researchers to introduce the term was de Bernis who put forward a number of new ideas. His work was close to the «State Monopoly Capitalism» thesis and was therefore profoundly influenced by a traditional variant of Marxism. In an article that appeared in 1975 in the revue Economique he proposed the substitution of the a priori identification of a general economic equilibrium with a hypothesis about the regulation of a system as a result of the interaction of its long term laws of functioning. He developed this idea by connecting it with the work of Perroux in a 1977 seminar on the idea of regulation in science (*L'idée de régulation dans les sciences*). Studying the relations between regulation or equilibrium in economic analysis (*Régulation ou équilibre dans l'analyse économique*) he demonstrated the significance of systems analysis for a modernization of non-orthodox economics. De Bernis explained that in capitalist societies essentially divergent forces are at work, but certain rules sometimes make these forces converge. His idea was that the things which play the same role as genetic codes in biology and allow convergence in economics are institutions. In this way it is possible to explain why, in certain periods if adequate institutions are established, a certain coherence, which can lead to economic growth, can be maintained in spite of the fact that capitalism is constituted and structured by forces that are essentially divergent. However the founding work of the regulation approach was published by Aglietta (1976). He introduced the concept of the fundamental «structural forms» of capitalist societies. These structural forms are at one and the same time economic and non-economic. When these forms reach a certain degree of coherence they ensure the regular development of the economy as a whole and allowing forces to converge and hence giving rise to the possibility of continuous growth over a relatively long period of time.

There is a terminological problem with the translation of the French term «régulation» into English. In French one distinguishes between the word «régulation» (a cybernetic term which indicates exactly the way the evolution of the elements of a system are adjusted to the functioning of the unit) and the word «réglementation» (group of institutionalised rules which can in effect serve the «régulation»). The English word regulation has more the second sense, but the English translations of «régulation» have only given the broadest sense of the French word.

The method of regulation and the accumulation regime first studied by this approach form the model of capitalist development dominant after the Second World War, «Fordism». Fordism is characterised (very schematically) by :
— a «Taylorist» division of work between activities of conception, skilled production, and unskilled production, all governed by hierarchical procedures;
— a system of accumulation based on the redistribution of the gains of productivity to the workers, in a way that will guarantee the growth of effective demand;
— a means of regulation that guarantees this redistribution through social legislation, collective agreements and Welfare States.

Since the first works of the School of Regulation translated into English were dedicated to the study of Fordism, the impression spread that «the theory of regulation is the theory of Fordism». This is untrue. The approach of regulation is a method that can be applied to many objects : models of economic development pre or post-Fordism, or even... theatre plays! The essential thing is to concern oneself with the relationship between a structure and its elements. Here we are going to concern ourselves with the relationship between a structured space and the agents within it, or with the relationship between global spaces and constituent sub-spaces.

As we have seen, theories of regulation find their origin and dynamism in an extensive culture, at the margins of the diverse disciplines of the social sciences, even if the economic remains dominants. This approach has suggested a new direction in research since the 1970s; it has also renewed conceptual frameworks and theoretical analysis. The regulationst paradigm continues to develop within an international association «Research and Regulation» (*Recherche et Régulation*)[1] under the presidency of Robert Boyer, and remains one of the best examples of global research in the social sciences, setting asid the constraints of disciplinary boundaries.

Spatial metaphors offered an inexhaustible reserve of images to the approach of regulation. And this is normal : as it was necessary to study the changes in time and modes of regulation and accumulation, these were themselves conceived as «synchronies» susceptible to spatial representations (cycles, stationary regimes). The reflection on the structuring of human geographical spaces became themselves fields of application of the approach once it was developed, whether they deal with the emergence or the reproduction of innovative territories or of forms of industrial organisation, or whether they deal with relations between local space and global space.

2 Theory of regulation and space

Preliminary work

In a preparatory work, still very structuralist (Althusserian even) on urban land rent, Lipietz (1974) began with a kind of spatial regime : the Economic and Social Division of Space, and asked himself by which mechanism this space reproduced itself or transformed itself under the private initiative of the property developers. He showed the role of the land prices and the institutions for urban planning, but underlined (especially Lipietz, 1975) two modes of regulation and production of urban space : competitive and monopolist, according to which the property developer or the public agency subjected or organised the modifications of the division of space resulting from multiple initiatives. This terminology was adopted as soon as the CEPREMAP (1977) report distinguished between the existence of many ways of regulation. In turn, the reflection on the prices of land, their active role and their divergence in relation to the Marxist «law of value», will clarify the distinction between the «esoteric» level (the world of structures) and the «exotérique» level (that of behaviour) (Lipietz 1983, 1984).

In the same way, the co-existence of different types of regions in the national space was conceived in Lipietz (1977) as the spatial spreading of the technological paradigm and of the regime of Fordist accumulation in the assembly lines. But already this very structuralist vision was qualified by the relative autonomy of the regions, and therefore raised the question of inter-regional regulation and the active role of regions in integrating themselves or not in the «inter-regional division of work». Once transposed to an international level, the approach of regulation was fully developed and this duality was underlined in Lipietz (1985a) : against the structuralist visions of dependency theories of the international division of labour that made the features of one region or country depend on its place in the global space, he outlined the fragility of the «international configurations», and the difficulties of their regulation, and above all the autonomy of spaces included, in this case the national States.

Following the same path, Aydalot (1984, 1986a) put forward the autoproduction of local territories... which would open the immense field of local regulation.

Research in the 1980s and 1990s industrial organisation, district, network, gouvernance, conventions,...

Following the breakthrough by Aydalot and by GREMI (who did not initially relate to the problem of regulation), the work which put forward the territorial analyses of innovation and economic and social organisation has experienced a revival since the end of the 1970s. The success and the growth of industrial regions would be due essentially to their internal dynamic. This vision broke as much with global structuralism (which can without doubt take into account the decline of regions : (Massey and Meegan, 1982) as with the theory of «stages of development» initiated by Rostow.

Schematically, one can distinguish many large categories of work. Initially there are those that one can group around the notion of industrial district, with three typical cases : the technological conglomerations (e.g. Silicon Valley), self-employed or small and medium sized entreprises (Third Italy), and financiers and services (the big metropoles). These works lean on the old intuition of Alfred Marshall, reinterpreted in accordance with the theory of transaction costs or the evolutionary theories of changing technique. The analyses tend to privilege the bargaining relations between firms, but under the influence of those who support regulation, the new research takes more and more into account the other forms of co-ordination between enterprises, the type of capital-labour relations prevailing over the territory, the politics of local development led by the elite etc...

At the beginning research was carried out in Italy. Giacomo Becattini (1992), Arnaldo Bagnasco and Carla Triglia, while working on «Third Italy», and insisting on the socially endogenous character of the development (the «social construction of the market»), analysed the type of industrial organisation of these regions, which presented themselves as a mixture of competition - emulation and co-operation at the heart of a system of small and medium sized firms. The method of regulation and the technological paradigm of this productive environment allowed for an old concept to be reintroduced : the «industrial district» described by Alfred Marshall in 1900, that is to say, the co-ordination, by the market and by reciprocity founded on geographic proximity, of a social division of work (vertical disintegration) between small firms specialising in a segment of the productive process.

Piore and Sabel (1984) interpreted the success of industrial districts as a particular case which is part of a much more general tendency. Referring to the approach of regulation, they only advanced to Fordist mass production, which is rigidly structured and which would be followed by a regime founded on flexible specialisation, the spatial form of which would be the district, just as the assembly line was a spatial form of the deployment of Fordism. This new industrial divide was attributed in effect on the one hand to the professionalism of the work force and on the other hand to decentralised innovation and co-ordination (by the market and reciprocity) between forms : two features of the social atmosphere of the industrial district.

In a similar way, and in interaction with these numerous influences, research was carried out in France on localised «industrial systems» by a team from Grenoble under the impulse of Courlet and Pecqueur (1991, 1992) while Gilly and his collaborators in Toulouse are working on the territorial aspect of industrialisation and innovation.

At the same time, the Californian school of economic geography represented essentially by Allen Scott, Michael Storper and Richard Walker, impressed by the growth of their state and particularly of Los Angeles, arrived at similar conclusions on a slightly different basis, indeed in extremely large centres, in which they later recognised patchworks of districts. Then, although knowing the approach of regulation from which they borrowed some terminology, they drew heavily from neo-Marxist or neo-classical analysis (those of Coase, 1937 and Williamson, 1975) on the dynamic of the division of labour and the external effects of agglomeration.

One saw also the emergence of a «Coase-Williamson-Scott paradigm», according to which industrial organisation arbitrated between the costs of internal organisation to the firm and the costs of transaction between firms. The agglomeration of firms in the same place minimises the costs of transaction. When the growing importance of economies of scope and the flexibility of their productive systems on the concentration in big firms is favourable to economies of scale, then Fordist spatial systems (vertically integrated) will disappear in the face of agglomerations of firms looking for minimal transaction costs (Capellin, 1988).

In the same way Strorper and Walker (1989) proposed a model of emergence of poles of growth which arose from nowhere in California. Allen Scott (1988a, b, 1993) progressively enriched his analysis of systems of contemporary production : social division of labour, transactions between the actors of a production space, different types of systemic relations, a Marshalian atmosphere and institutions, role of political or almost political agents ...

One specific form of districts was studied by economists, geographers and sociologists, under the name of «technopoles», spaces created by a voluntary industrial policy (Japan, Germany, France) or as a first stage for moving towards flexible accumulation (Benko, 1991; Scott, 1993; Castells, Hall, 1994).

From the smallest Italian district to world-wide megapoles, the new technological paradigm of «flexible specialisation» would boost not only the return of factories and offices towards urban areas, but more so the revival of the quantitative growth of the metropoles : a spatial form found at the end of the Fordist crisis. The future hierarchy of world wide urban towns and regions would result from the internal strategy of these districts or collections of districts.

At the cross-roads of reflection on districts and evolutionary approaches to the diffusion of technical change, another approach considers the territory itself as an «innovative environment». This possibility was particularly developed by the European team GREMI (Group for European Research on Innovative Environments - Association Philippe Aydalot). At the outset, one asks oneself what external conditions are necessary for the birth of the firm and for the adoption of innovation. The researchers considered that the firm did not exist before the local milieux, but that it was exuded by them. (Aydalot, 1986). One seeks to theorise about the different forms of interdependence which weave themselves into the territory and intervene in the technological development, in incorporating very different elements. Recent studies recuperate the work carried out in the domain of the theory of industrial organisation, and the analysis of industrial districts (Maillat and Perrin, 1992). They return more and more explicitly to the approaches, that have revolutionised the approach of the emergence and diffusion of innovation. Taking the opposite path, the evolutionists have put emphasis more and more on the «environment», and therefore on territory favourable to innovation (Dosi and Salvatore, 1992).

Facing this work that everyone was theorising about as a canonical form of «post-Fordism», borrowing some elements from the approach of regulation, Danièle Leborgne and Alain Lipietz (1988) attempted to develop a framework using this approach more systematically. This first reflection was completed by two other essays (1991, 1992).

In this widely translated triptych, they underlined many points.

*Territories are not geared towards a «flexibility» of the wage contract. Some yes, but others privilege the negotiated involvement of the workers and their qualifications, and this option is contradictory to a very great flexibility.

*The tendency towards «vertical disintegration» of the processes of production is in effect general, but there exist very different forms of co-operation and hierarchy between the firms.

*The territories or the capital-labour relationship are the most flexible, often presenting quite «brutal» market relations between the firms themselves; contrarily those where «fidelity» between capital and labour prevail, often see forms of partnership prevail between the firms.

*These diverse forms of evolution of territories (among which the authors propose a bringing together of the different types of localised productive systems proposed by geographers) correspond with the «defensive» or «offensive» strategies of their elites.

One can see : the French who favour regulation remain sceptical about the uniqueness of «post-Fordism», flexible accumulation, and its spatial expression (the district). On the other hand, Belgian economists and geographers who work according to the same methodology, advance a more united vision of post-Fordism : Moulaert and Swyngedouw (1988, 1992).

The remarks of Leborgne and Lipietz on the variety of forms of inter-firm relations strongly drew attention to the relevance of the model of «Marshalian districts», regulated by the combination of market relations and an «atmosphere» of reciprocity. Economists, geographers and engineers went on to examine more attentively the forms of organisation of relations between units of production and their spatial deployment : the hierarchical form of the assembly line, and the purely market form of the district of a «Coase-Williamson-Scott-type» are only two extreme and caricatured forms.

We will call «network» the spatial dimension of one form of regulation of relations between productive unities, and «governing» the mode of regulation of these relations, which is in general the combination of different forms : hierarchy, subcontracting, partnership, «atmosphere», public agencies or para-public. Storper and Harrison (1992) show the great variety in ways of gouvernance. Veltz (1990, 1992), an engineer, worked on the systems of big firms and the territorial hierarchy of production. The logic of externalisation does not necessarily signify the return of the market. The hierarchical organisation transforms itself into a network organisation, built around notions of partnership, contractual relations, flexibility, and inter-firm co-operation. The same terms appeared of course in the analysis of districts and environments.

Storper and Harrison (1992), we have already seen, introduced into spatial analysis the notion of «gouvernance», as a form of inter-firm organisation, going beyond market relations. It is a question of a complex theorisation resulting form increased reflections in the domains of industrial organisation, the division of work, the institutions, the conventions, and the possible location. But it is understood that this concept can extent to every system of human territorial relations.

Salais and Storper (1993) analyse the possible forms of economic co-ordination between people, products, conventions, elementary action registers and the forms of uncertainty to which are compared economic actors, and the «possible worlds» of production in this way envisaged are compared with real worlds, through regional empirical studies. Salais and Storper construct also national economic identities for many countries.

In this way a broad sense of the concept of gouvernance emerges : the collective ways of regulation between the pure market and pure politics (of the Nation-State), that is to say what Gramsci called «civil society». The School of Regulation owes it seems a great deal to Italy.

Earlier, it was argued that those who favour regulation took up the problem of the articulation of local spaces and global spaces. It even served as a paradigm for the modal reflection on the regulation approach : the individual social relationship processes of Fordist production tended however to establish itself. Lipietz (1985, 1986) was in complete agreement : there is indeed a «world economy», but it hasn't a causal force. Multinational firms are not any longer the creators of the «New international division of labour», and this division is rather a configuration, an encounter, weakly regulated by national trajectories, certain of which, present a strong dependence on the global context (whence their names of primitive Taylorisation, peripheral Fordism, etc...)

This notion of a vague regime and of weak regulation confirmed the point of view of the specialists of «international regimes» like Krasner (1982) and Koehane (1982), and more generally the School of Cornell and the *International Organisation* review. But it also opened a new way to reform «neo-structuralism» that could resolve itself to apprehend the territories (they were innovative districts) like monads exchanging products.

The debate caused a storm among geographers, sociologists and regional economists who were divided into two camps that drew on different versions of the regulationist heritage : those who privileged the structure of the local; and those who privileged the constraints of the global and were critical of the «mythical geography of flexible accumulation» (Amin and Robins, 1992) and underlined the weight of oligopolistic structures (Martinelli and Schoenberger, 1992) and the dominant megapolises (Veltz, 1992). Three collections develop this debate (Storper and Scott, 1992; Benko and Dunford, 1991; and in French Benko and Lipietz, 1992).

It must be admitted however that behind this debate (local/global) is hidden a certain incapacity on the part of those who favour regulation themselves to identify the traits of post-Fordism. The weight of the hypothesis on flexible specialisation in the area of the influence of regulationism in the midst of the Anglo-Saxon world (see for example the journal *Society and Space*) has often clouded the issue taking too literally the hypothesis (that originated in the earlier work of Mistral) that the transition of an area from Fordism to post-Fordism implied the adoption of a single new model of development. French doubts about the singularity of post-Fordism (Boyer, 1992; Leborgne and Lipietz, 1992) re-opened the debate by asking if different models of development coexisted in a single global arena.

3 Conclusion and research directions

As has just been shown, the richness of the regulation approach to the analysis of space leaves, for the moment, more questions unanswered than answered, all the more so because «space» is by nature an interdisciplinary terrain.

It is first of all the «regulation of the local» and the concept of governance that it should develop. There is also the question of interlocking spaces. Between the resurgence of the «local» (as a condition of competition and of social regulation) and of globalization as the space in which the economy and, indeed, culture develop, the regulation approach allows for a profoundly renewed concept of international relations. Some explicitly regulationary texts (Lipietz, 1992, 1993, Leborgne and Lipietz, 1989, 1991, 1992a) took on board the question as such : the influence of international modes of regulation on the selection of technological paradigms, the macro economy of continental agreements (CEE, ALENA), and the possibility of the coexistence of different models of development in the same area of free-trade. These reflections rejoin the evolution of «neo-structuralists» (Palan and Gills, 1994) and are manifest in the launching of the *Review of International Political Economy*.

Finally, it is possible to observe a certain «rapprochement» between the Anglo-Saxon sociological (and even aesthetic) reflections on space, inspired by Giddens (1984), his «structuration theory» and the debate about «structure versus agency» (Lazar, 1992) and the problematic of regulation. This convergence is further accelerated by the even more ambitious bringing together by Harvey (1989) of «post-Fordism» and «post-modernity» (in architecture and urban planning). Works such as those of Gregory and Urry (1985) mark out this convergence which makes a link between the French approach of regulation and radical Anglo-Saxon geography (represented for example by *Antipode*) and even with Feminist geography, which poses the dual question of the spatial deployment of relations of gender and the social (and hence localised) construction of gender (MacDowell, 1992).

In short one can say that it is certainly a difficult and very vast but interesting programme of research. I will put forward three questions which seem to me to be fundamental : the status of the actor, the methodology of research and the process of accumulation. One could imagine producing an analysis of the dynamic origin of institutions, and of their creation by a «structuring interaction» that conforms with a broadened rationality. It is interesting in this framework of the theory of regulation to think in terms of hybridisation, where there is a combination of the new and the old, the local and the global and the social and the economic. There are so many areas to follow in research in regulation during the coming years.

Note :

1 The «Research and Regulation» (Recherche et Régulation) association was founded in 1994. It is led by Robert Boyer, and groups together the main researchers in the world and in different disciplines who are interested in the theory of regulation. The address is the following : Prof. Yves Saillard, Executive Secretary, IREPD, BP.47, 38040 Grenoble - St. Martin d'Héres, Cedex 9, France. The association publishes La lettre de la Régulation.

References

Aglietta, M., 1976, Régulation et crises du capitalisme, Paris: Calmann-Lévy

Aglietta, M., 1979, A theory of capitalist regulation, London: New Left Books

Amin, A. and Robins, K., 1992, 'Le retour des économies régionales? La géographie mythique de l'accumulation flexible', in Benko, G. B. and Lipietz, A., eds., Les régions qui gagnent. Districts et réseaux: les nouveaux paradigmes de la géographie économique, Paris: PUF, 123-161

Aydalot, P., 1980, Dynamique spatiale et développement inégal, Paris: Economica

Aydalot, P., ed., 1984, Crise et espace, Paris: Economica

Aydalot, P., ed., 1986a, Milieux innovateurs en Europe, Paris: GREMI

Aydalot, P., 1986b, Les technologies nouvelles et les formes actuelles de la division spatiale du travail, Paris, Dossier du Centre Economie, Espace, Environnement, n° 47

Aydalot, P. and Keeble, D., eds., 1988, High Technology Industry and Innovative Environments: The European Experience, London: Routledge

Bagnasco, A., 1981, 'Labour market, class structure and regional formations in Italy', International Journal of Urban and Regional Research, 5, 1, 40-44

Bagnasco, A. and Trigilia, C., 1988, 1993, La construction sociale du marché. Le défi de la troisième Italie, Cachan: Ed. de l'ENS-Cachan

Baslé, M., Baulant, C. and al., 1988, Histoire des pensées économiques, Paris: Dalloz/Sirey

Becattini, G., 1990, 'The Marshallian industrial district as a socio-economic notion', in Pyke, F., Becattini, G. and Sengenberger, W., eds., Industrial Districts and Inter-firm Co-operation in Italy , Genève: ILO, 37-51

Becattini, G., 1991, 'Le district industriel: milieu créatif', Espaces et Sociétés, 66/67, 147-163

Becattini, G., 1992, 'Le district marshallien: une notion socio-économique', in Benko, G. B. and Lipietz, A., eds., Les régions qui gagnent. Districts et réseaux: les nouveaux paradigmes de la géographie économique, Paris: PUF, 35-55

Benko, G., ed., 1990, *La dynamique spatiale de l'économie contemporaine*, La Garenne-Colombes: Editions de l'Espace Européen

Benko, G., 1991, *Géographie des technopôles*, Paris: Masson

Benko, G. and Dunford, M., eds., 1991, *Industrial change and regional development: the transformation of new industrial spaces*, London: Belhaven/Pinter

Benko, G. and Lipietz A., eds., 1992, *Les régions qui gagnent*, Paris: PUF

Boltanski, L. and Thévenot, L., 1987, *Les économie de la grandeur*, Paris: PUF

Boltanski, L. and Thévenot, L., 1991, *De la justification. Les économie de la grandeur*, Paris: Gallimard

Bourdieu, P., 1980, *Question de sociologie*, Paris: Minuit

Bowles, S. and Gintis, H., 1993, 'The Revenge of Homo Economicus: Contested Exchange and the Revival of Political Economy' *Journal of Economc Perspectives*, 7, 1, 83-102

Bowles, S., Gordon, D. M. and Weisskopf, T. E., 1983, *Beyond the Waste Land*, New York: Doubleday

Boyer, R., 1978, 'Les salaires en longue période', *Economie et Statistique*, 103 (September) 27-57

Boyer, R., 1986, *La théorie de la régulation: une analyse critique*, Paris: Editions La Découverte

Boyer, R., ed., 1988, *The search for labour market flexibility.The European economies in transition*, Oxford: Clarendon Press

Boyer, R., 1992, 'Les alternatives au fordisme. Des années 1980 au XXIe siècle', *in* Benko, G. B. and Lipietz, A., eds., *Les régions qui gagnent. Districts et réseaux: les nouveaux paradigmes de la géographie économique*, Paris: PUF, 189-223

Boyer, R., 1995, 'La théorie de la régulation dans les annéess 1990', *Actuel Marx*, 17, 19-38

Boyer, R. and Durant, J-P., 1993, *L'après-fordisme*, Paris: Syros

Boyer, R. and Saillard, Y., eds., 1995, *Théorie de la régulation: l'état des savoirs*, Paris: La Découverte

Castells, M. and Hall, P., 1994, *Technopoles of the World. The making of 21st Century Industrial Complexes*, London: Routledge

CEPREMAP, 1977, *Approches de l'inflation: l'exemple français*, Paris: CEPREMAP, Rapport au CORDES, miméo

Coase, R. H., 1937, 'The nature of the firm', *Economica*, 4, 16, 386-405

Colletis, G., Courlet, C. and Pecqueur, B., 1990, *Les systèmes industriels localisés en Europe*, Grenoble, IREPD

Colletis, G., Pecqueur, B., 1993, Intégration des espaces et quasi-intégration des firmes: vers de nouvelles rencontres productives?, *Revue d'Economie Régionale et Urbaine*, 3, 489-508

Commons, J. R., 1934, *Institutional Economics*, New York: Macmillan

Coriat, B., 1982, *L'atelier et le chronomètre*, Paris: Christian Bourgois.

Coriat, B., 1991, *Penser à l'envers*, Paris: C. Bourgois

Courlet, C. and Pecqueur, B., 1992, 'Les système industriel localisés en France: un nouveaux modèle de développement', *in* Benko, G. B. and Lipietz, A., eds., *Les régions qui gagnent. Districts et réseaux: les nouveaux paradigmes de la géographie économique*, Paris: PUF, 81-102

Courlet, C. and Pecqueur, B., 1991, 'Systèmes locaux d'entreprises et externalités: un essai de typologie', *Revue d'Economie Régionale et Urbaine*, 3/4, 391-406

Courlet, C., Pecqueur, B. and Soulage, B., 1993, 'Industrie et dynamiques de territoires', *Revue d'Economie Industrielle*, 64, 7-21

Courlet, C. and Soulage, B., eds., 1994, *Industrie, territoires et politiques publiques*, Paris: L'Harmattan

De Bernis, G., 1977, 'Régulation ou équilibre dans l'analyse économique' *in* Lichnerowicz, A., Perroux, F. and Gadoffre, P., eds., *L'idée de régulation dans les sciences*, Paris: Maloine éditeur, 85-101

Dosi, G., Freeman, C., Nelson, R., Silverberg, G. and Soete, L., 1988, *Technical Change and Economic Theory*, London: Pinter

Dosi, G. and Salvatore, R., 1992, 'The Structure of Industrial Production and the Boundaries Between Firms and Markets', *in* Storper, M. and Scott, A. J., eds., *Pathways to Industrialization and Regional Development*, London: Routledge, 171-192

Duménil, G., Glick, M. and Rangel, J., 1987, 'Theories of the Great Depression: why did profitability matter?', *Review of Radical Political Economics*, 19, 2, 16-42

Dunford, M, 1988, *Capital, the state and regional development*, London: Pion.

Dunford, M., 1990, 'Theories of regulation', *Environment and Planing D: Society and Space*, 8, 3, 297-321

Dupuy, C. and Gilly, J-P., eds., 1993, *Industrie et territoires en France. Dix ans de décentralisation*, Paris: La Documentation Française

Durant, J-P., ed., 1993, *Vers un nouveaux modèle productif*, Paris: Syros

Favereau, O., 1995, 'L'économie des conventions', *Actuel Marx*, 17, 103-114

Freeman, C., 1987, *Technology policy and economic performance: lessons from Japan*, Oxford: Frances Pinter

Freeman, C., 1988, 'The factory of the future: the productivity paradox, Japanese just-in-time and information technology' *PICT Policy Research Papers*, 3 (May) London: Economic and Social Research Council

Freeman, C., Clark, J. and Soete, L., 1982, *Unemployment and Technical Innovation*, London: Pinter

Freeman, C. and Perez, C., 1986, *The Diffusion of Technical Innovations and Changes of Technico-Economic Paradigm*, Falmer: University of Sussex, SPRU

Giddens, A., 1984, *The Constitution of Society*, Cambridge: Polity Press

Gilly, J-P. and Grossetti, M., 1993, 'Organisation, individus et territoires. Le cas des systèmes locaux d'innovation', *Revue d'Economie Régionale et Urbaine*, 3, 449-468

Glyn, A., Hughes, A., Lipietz, A. and Singh, A., 1988, *The rise and fall of the Golden Age*, Cambridge: Department of Applied Economics Discussion Paper, 884

Gregory, D., Urry, J., eds., 1985, *Social Relations and Spatial Structures*, London: Macmillan

Harvey, D., 1989, *The Condition of Postmodernity*, Oxford: Basil Blackwell

Hirsch, J. and Roth, R., 1987, *Das neue Gesicht des Kapitalismus.Vom Fordismus zum Post-Fordismus*, Hamburg: VSA Verlag

Hirsch, J., 1983, 'Fordist security state and new social movements', *Kapitalistate*, 10/11, 75-88.

Hirsch, J., 1984, 'Notes towards a reformulation of state theory', *in* Haenninen, S. and Paldan, L., eds., *Rethinking Marx*, Berlin: Argument Verlag, 155-60

Inoué, Y. and Yamada, T., 1995, 'Japon. Démythifier la régulation', *in* Boyer, R. and Saillard, Y., eds., *Théorie de la régulation: l'état des savoirs*, Paris: La Découverte, 408-416

Jessop, B., 1983, 'Accumulation strategies, state forms, and hegemonic projects', *Kapitalistate*, 10, 89-111

Jessop, B., 1990a, 'Regulation Theories in Retrospect and Prospect', *Economy and Society*, 19, 2, 153-216

Jessop, B., 1990b, *State Theory: Putting Capitalist States in their Place*, Cambridge: Polity Press

Keohane, R., 1982, 'The demand for international regimes', *International Organization*, 36, 2

Krasner, S., 1982, 'Regimes and the limits of realism: Regimes as autonomous variables', *International Organization*, 36, 2

Lash, S. and Urry, J., 1994, *Economies of Signs and Space*, London: Sage

Lazar, J., 1992, 'La compétence des acteurs dans la théorie de structuration de Giddens', *Cahiers Internationaux de Sociologie*, 39, 399-416

Leborgne, D. and Lipietz, A., 1988a, 'L'après-fordisme et son espace', *Les Temps Modernes*, vol 43, n°501 (avril), 75-114

Leborgne, D. and Lipietz, A., 1988b, 'New technologies, new modes of regulation: some spatial implications', *Environment and Planing D: Society and Space*, 6, 3, 263-280

Leborgne, D. and Lipietz, A., 1990, 'Pour éviter l'Europe à deux vitesses', *Travail et Société*, 15, 2, 189-210

Leborgne, D. and Lipietz, A., 1991, 'Idées fausses et questions ouvertes de l'après-fordisme', *Espaces et Sociétés*, 66/67, 39-68

Leborgne, D. and Lipietz, A., 1992a, 'Flexibilité offensive, flexibilité défensive. Deux stratégies sociales dans la production des nouveaux espaces économiques', *in* Benko, G. B. and Lipietz, A., eds., *Les régions qui gagnent. Districts et réseaux: les nouveaux paradigmes de la géographie économique*, Paris: PUF, 347-377

Leborgne, D. and Lipietz, A., 1992b, 'Conceptual Fallacies and Open Questions on Post-Fordism', *in* Storper, M. and Scott, A. J., eds., *Pathways to Industrialization and Regional Development*, London: Routledge, 332-348

Lecoq, B., 1991, 'Organisation industrielle, organisation territoriale: une approche intégrée sur le concept de réseau', *Revue d'Economie Régionale et Urbaine*, 3/4, 321-341

Lecoq, B., 1993, 'Proximité et rationalité économique', *Revue d'Economie Régionale et Urbaine*, 3, 469-486

Lewis, D. K., 1969, *Convention : A Philosophical Study*, Cambridge, Mass.: Harvard University Press

Lipietz, A., 1974, *Le tribut foncier urbain*, Paris: Maspero

Lipietz, A., 1977, *Le capital et son espace*, Paris: Maspéro

Lipietz, A., 1979, *Crise et inflation. Pourquoi?* Paris: Maspero

Lipietz, A., 1983, 1985, *The enchanted world. Inflation, credit and the world crisis*, London: Verso

Lipietz, A., 1984a, 'A Marxist Approach to Urban Ground Rent', *in* Ball, ed., *Land Rent, Housing and the Planning System*, London: Croom Helm

Lipietz, A., 1984b, 'De la nouvelle division du travail à la crise du fordisme périphérique', *Espaces et Sociétés*, 44, 51-78

Lipietz, A., 1985, *L'audace ou l'enlisement*, Paris: La Découverte

Lipietz, A., 1986a, 1987, *Mirages et miracles. The crises of global Fordism*, London: Verso

Lipietz, A., 1986b, 'New tendencies in the international division of labor: regimes of accumulation and modes of regulation', *in* Scott, A. J and, Storper, M., eds., *Production, Work, Territory. The geographical Anatomy of Industrial Capitalism*, London: Allen and Unwin, 16-40

Lipietz, A., 1987, 'La régulation: les mots et les choses', *Revue Economique*, 38, 5, 1049-1060

Lipietz, A., 1989, *Choisir l'audace*, Paris: La Découverte

Lipietz, A., 1990a, 'Après-fordisme et démocratie', *Les Temps Modernes*, 45, 524, (mars) 97-121

Lipietz, A., 1990b, 'La trame, la chaîne et la régulation: un outil pour les sciences sociales', *Economies et Sociétés, Cahiers de l'ISMEA*, Série R-5, 24, 12, 137-174

Lipietz, A., 1990c, 'Le national et le régional: quelle autonomie face à la crise capitaliste mondiale?', *in* Benko, G. B., ed., *La dynamique spatiale de l'économie contemporaine*, La Garenne-Colombes: Editions de l'Espace Européen, 71-103

Lipietz, A., 1992, *Berlin, Bagdad, Rio*, Paris: Quai Voltaire

Lipietz, A, 1993, *Vert espérance*, Paris: La Découverte

Lipietz, A., 1994, 'De l'approche de la régulation à l'écologie politique : une mise en perspective historique', *Futur Antérieur*, (Numéro spécial hors série), 71-99

Lipietz, A., 1995a, 'Ecologie politique régulationiste ou économie de l'environnement ?', *in* Boyer, R. and Saillard, Y., eds., *Théorie de la régulation : l'état des savoirs*, Paris: La Découverte, 350-356

Lipietz, A., 1995b, 'De la régulation aux conventions: le grand bond en arrière?', *Actuel Marx*, 17, 39-48

Lyon-Caen, A. and Jeammaud, A., 1986, *Droit du travail, démocratie et crise*, Arles: Actes Sud

Maillat, D., 1992, 'Milieux et dynamique territoriale de l'innovation', *Revue Canadiennes des Sciences Régionales / Canadian Journal of Regional Science*, 15, 2, 199-218

Maillat, D., Crevoisier, O. and Lecoq, B., 1991, 'Réseaux d'innovation et dynamique territoriale', *Revue d'Economie Régionale et Urbaine*, 3/4, 407-432

Maillat, D. and Perrin, J-C., eds., 1992, *Entreprises innovatrices et développement territorial,* Neuchâtel: EDES

Maillat, D., Quevit, M. and Senn, L., eds., 1993, *Réseaux d'innovation et milieux innovateur: un pari pour le développement régional,* Neuchâtel: EDES

Martinelli, F. and Schoenberger, E., 1992, Les oligopoles se portent bien, merci! Eléments de réflexion sur l'accumulation flexible, *in* Benko, G. B. and Lipietz, A., eds., *Les régions qui gagnent. Districts et réseaux: les nouveaux paradigmes de la géographie économique,* Paris: PUF, 163-188

Massey, D., 1984, *Spatial Division of Labour. Social Structures and the Geography of Production,* London: Macmillan

Massey, D. and Meegan, R., 1982, *The Anatomy of Job Loss. The how, why and where of employment decline,* London: Methuen

McDowell, L., 1993, 'Doing gender: feminism, feminists, and research methods in human geography,' *Transactions, Institut of British Geographers,* 17, 399-416

Mistral, J., 1986, 'Régime international et trajectoires nationales', *in* Boyer, R., ed. *Capitalismes fin de siècle,* Paris: PUF, 167-201

Mitchell, W. C., 1913, *Business Cycles,* New York

Moulaert, F. and Swyngedouw, E. A., 1988, 'Développement régional et géographie de la production flexible', *Cahiers Lillois d'Economie et de Sociologie,* 11, 81-95.

Moulaert, F. and Swyngedouw, E. A., 1989, 'A regulation approach to the geography of flexible production systems', *Environment and Planning D: Society and Space,* 7, 3, 327-345.

Moulaert, F., Swyngedouw,. A. and Wilson, P., 1988, 'Spatial Responses to Fordist and Post-Fordist Accumulation and Regulation', *Papers of the Regional Science Association,* 64, 11-23

Moulaert, F. and Swyngedouw, E., 1992, 'Accumulation and organization in computing and communications industries: a regulationist approach', *in* Cooke, P , Moulaert, F. and al. *Towards Global Localisation. The Computing and Telecomminications Industries in Britain and France,* London: UCL Press, 39-60

Nelson, R., ed., 1993, *National Innovation Systems,* Oxford: Oxford University Press

Nelson, R. and Winter S., 1982, *An Evolutionary Theory of Economic Change,* Cambridge, Mass.: The Belknap Press of Harvard university Press

Ominami, C., 1986, *Le tiers monde dans la crise,* Paris: La Découverte

Orléan, A., ed. 1994, *Analyse économique des conventions,* Paris: PUF

Palan, R. and Gills, B., 1994, *Transcending the State-Global Divide,* London: Lynne Rienner Publishers

Peck, J. and Miyamachi, Y., 1994, 'Regulation Japan? Regulation theory versus the Japanese experience', *Environment and Planning D: Society and Space,* 12, 6, 639-674

Perrin, J. C., 1990, 'Organisation industrielle: la composante territoriale', *Revue d'Economie Industrielle,* 51, 276-303

Perrin, J. C., 1991, 'Réseaux d'innovation — milieux innovateurs. Développement territorial', *Revue d'Economie Régionale et Urbaine,* 3/4, 343-374

Perrin, J. C., 1992, 'Pour une révision de la science régionale. L'approche par les milieux', *Revue Canadiennes des Sciences Régionales / Canadian Journal of Regional Science,* 15, 2, 155-197

Piore, M. and Sabel, C. F., 1984, *The second industrial divide: possibilities for prosperity,* New York: Basic Books

Salais, R. and Storper, M., 1993, *Les mondes de production. Enquête sur l'identité économique de la France,* Paris: Ed. de l'EHESS

Schmoller G., 1874, 1902, Politique sociale et politique économique: questions fondamentales, Paris: Giard et Brière

Scott, A. J., 1988a, *Metropolis, From the Division of Labor to Urban Form,* Los Angeles: University of California Press

Scott, A. J., 1988b, *New Industrial Spaces,* London: Pion

Scott, A. J., 1993, *Technopolis. High-Technology Industry and Regional Development in Southern California,* Berkeley, CA.: University of California Press

Scott, A. J. and Storper, M., 1991, Le développement régional reconsidéré, *Espaces et Sociétés,* 66/67, 7-38

Scott, A. J. and Storper, M., 1992, 'Industrialisation and Regional Development', *in* Storper, M. and Scott, A. J., eds., *Pathways to Industrialization and Regional Development,* London: Routledge, 3-17

Scott, A. J. and Storper, M., eds., 1986a, *Production, Work, Territory. The geographical Anatomy of Industrial Capitalism,* London: Allen and Unwin

Storper, M. and Harrison, B., 1992, 'Flexibilité, hiérarchie et développement régional: les changements de structure des systèmes productifs industriels et leurs nouveaux modes de gouvernance dans les années 1990', *in* Benko, G. B. and Lipietz, A., eds., *Les régions qui gagnent. Districts et réseaux: les nouveaux paradigmes de la géographie économique,* Paris: PUF, 265-291

Storper, M. and Scott, A. J., 1989, 'The geographical foundations and social regulation of flexible production complexes', *in* Wolch, J. and Dear, M., eds., *The Power of Geography,* London: Unwin & Hyman, 21-40

Storper, M. and Scott, A. J., eds., 1992, *Pathways to Industrialization and Regional Development,* London: Routledge

Storper, M. and Walker, R., 1989, *The Capitalist Imperative. Territory, Technology and Industrial Growth,* Oxford: Basil Blackwell

Storper, M. and Scott, A. J., eds., 1992, *Pathways to Industrialization and Regional Development,* London: Routledge

Swyngedouw, E. A., 1992, 'Territorial organisation and the space / technology nexus', *Transactions of the Institute of British Geographers,* 17, 4, 417-433

Swyngedouw, E. A. and Kesteloot, C., 1988, 'Le passage sociospatial du fordisme à la flexibilité: une interprétation des aspects spatiaux de la crise et de son issue', *Espaces et Sociétés,* 54/55, 243-262

Swyngedouw, E. A., 1988, 'The geography of high-technology production in France and the technology/defense nexus', *L'Espace Géographique,* 17, 4, 269-276.

Veblen, T., 1899, 1953, *The theory of the leisure class: an economic study of institutions,* New York: Mentor Books

Veltz, P., 1990, 'Nouveaux modèles d'organisation de la production et tendances de l'économie territoriale', *in* Benko, G., ed., *La dynamique spatiale de l'économie contemporaine,* La Garenne-Colombes: Editions de l'Espace Européen, 53-69

Veltz, P., 1992, 'Hiérarchies et réseaux dans l'organisation de la production et du territoire', *in* Benko, G. and Lipietz, A., eds., *Les régions qui gagnent. Districts et réseaux: les nouveaux paradigmes de la géographie économique,* Paris: PUF, 293-313

Wagner, A., 1876, 1904, *Les fondements de l'économie politique,* Paris: Giard et Brière

Weiller, J. and Carrier, B., 1994, *L'économie non conformiste en France au XXe siècle,* Paris: PUF

Williamson, O. E., 1975, *Markets and Hierarchies. Analysis and Antitrust Implications,* New York: Free Press

Georges Benko
Université de Paris I – Panthéon-Sorbonne
191, rue Saint-Jaques
75005 Paris
France

PART V

POLITICS

14. Territoriality and the state

Ron J. Johnston

Territoriality is a concept that has been used in discussions of the nature of behaviour in a variety of social a natural sciences. As Sack (1983) pointed out in his important theoretical essay on human territoriality, some writers - notably those drawing on certain ethological work (e.g. Ardrey, 1969) - present it as a biological drive or aggressive instinct, a genetically transmitted behavioural trait which humans share with many other animal species. He rejects that view, however, preferring to present territoriality, as defined in his later book (Sack, 1986: 2), as «a human strategy to affect, influence and control». This essay shares Sack's rejection of the biological case, and extends his argument with particular reference to the geography of the state.

The definition of territoriality used here is provided by Smith (1986a: 482) :

> The attempt by an individual or group to influence or establish control over a clearly demarcated territory which is made distinctive and considered at least partially exclusive by its inhabitants or those who define its bounds.

In this, he closely follows Sack in emphasising the importance of power as a human relationship the exercise of which territoriality is used as an important strategy : Sack's definition (1986: 19) is that territoriality is

> the attempt by an individual or group to affect, influence or control people, phenomena and relationships, by delimiting and asserting control over a geographic area.

But Smith goes further, noting that people identify the use of territoriality strategies as based in needs for «identity, defence and stimulation» but stressing that those needs are socially produced not biologically given, so that use of the strategy is «conditioned primarily by cultural norms and values which vary in structure and function from society to society, from one time period to another,

213

G. B. Benko and U. Strohmayer (eds.), Geography, History and Social Sciences, 213–225.
© 1995 *Kluwer Academic Publishers. Printed in the Netherlands.*

and in accordance with the scale of social activity». At the societal level, he argues

> territoriality becomes a means of regulating social interaction and a focus and symbol of group membership and identity, ranging from the scale of urban gangs and their turf, through patterns of territorial regionalism, to the compartmentalization of the world into a system of states.

The later sections of this essay focus on the latter scale. First, however, Sack's important seminal contribution is outlined in greater detail.

1 Sack's theory of human territoriality

As already stressed, Sack's theory is non-deterministic, as emphasised in his statement (1983: 57) that :

> under certain conditions territoriality is a more effective means of establishing differential access to people, or resources, than is non-territoriality

and he suggests ten reasons why this is so. Those ten reasons (he calls them «tendencies») refer to two major characteristics of territory: it can be bounded, and thus readily communicated; and it can be used to displace personal relationships, between controlled and controller, by relationships between people and «the law of the place» - territory is thereby reified in order to promote personal ambitions impersonally. He then sets out a variety of combinations of the ten tendencies which illustrate the various uses of a territorial strategy: for example, use of a territorial strategy, such as splitting up a factory into a series of separate workshops the members of which are in competition, facilitates a policy of «divide and conquer».

Sack ends his seminal introductory essay by pointing to the criteria that must be met if his general theory is to be vindicated. A large number of cases must be studied, he argues: in order to assess whether the adduced reasons for territorial rather than nonterritorial behaviour are valid; to determine the precise conditions under which the operation of a territoriality strategy becomes and advantage; determining whether there are conditions under which territoriality is a necessary strategy; and identifying the degree of advantage that a territorial strategy can bring to an organisation. His later book (Sack, 1986) undertakes this, at least in part. An initial essay looks at the increasing use of territoriality over time, and this is followed by detailed analyses of: the organisation of the Roman Catholic Church; the territorial organisation of local government in the United States; and the spatial structuring of workplaces. (In the 1983 essay, he used the example of the organisation of the United States' Army). Although his book is not set out explicitly to assess his arguments against the criteria set out at the end of the 1983 essay, there is no doubt that Sack's conclusions are almost entirely positive with regard to his theory's potential. Thus he defines territoriality as (1986: 216) :

> the basic geographic expression of influence and power... an essential link between society, space and time. Territoriality is the backcloth of geographical context - it is the device through which people construct and maintain spatial organizations... [it] is not an instinct or drive, but rather a complex strategy to affect, influence, and control access to people, things and relationships.

He doesn't identify exactly the conditions in which territoriality will be used as a strategy, but does implicitly conclude that it is a necessary one (p. 219) :

> Whatever the goals of a society may be... and whatever the geographical scale... a society, simply as a complex organization, will need territoriality to coordinate efforts, specify responsibilities, and prevent people from getting in each other's way. And since territoriality of one sort or another will likely be employed, we must be aware... [that it] is not only a means of creating and maintaining order, but is a device to create and maintain much of the geographic context through which we experience the world and give it meaning.

That final sentence provides the basis for extending Sack's treatment in the remainder of this essay.

2 The state as territorial container

One of Sack's examples of the use of territoriality is on the American system of local government, in which he concluded that (1986: 167) :

> The development of hierarchies of territories has defined communities of contain, channel, and mold the geographically dynamic processes; has heightened the effect of impersonality; and, in many cases, has increased the bureaucratization and centralization of power.

Much of the discussion focuses on the provision of «public goods», for the provision of which he believes that «Territorial hierarchy is...thought to be unavoidable in complex capitalist political systems». He does not, however, address many of the issues relating to the nature and operation of the state, which have concerned theorists in recent years. States are widely recognised as territorially defined institutions. Dunleavy and O'Leary (1987, for example, define a modern state with regard to five characteristics, which include statements that (p. 2) :

> The state is sovereign... within its territory

and

> The state's sovereignty extends to all the individuals within a given territory.

But the relationship between a state's sovereignty and its territory is not discussed (territory and territoriality are not listed in the book's index), a feature shared with other texts on the nature of the state (e.g. Alford and Friedland, 1985; King, 1986), including one by two geographers (Clark and Dear, 1984; for an elaboration of this point, see Johnston, 1990a).

Two exceptions to the general tendencies set out in the previous paragraph are the works of social theorists Giddens and Mann. Giddens, in *The Nation-State and Violence* (1985), defines the state as «a political organization whose rule is territorially ordered and which is able to mobilize the means of violence to sustain that rule» (p. 20), and argues that the growth of the absolutist state, with its much wider span of power than its feudal predecessor, required a clearer definition of territorial structures. With the replacement of absolutist by modern nation-states territory became even more important; a nation-state is defined (p. 120) as «a bordered power-container... the pre-eminent power-container of the modern era».

A major activity within nation-states (by other agencies as well as by the state apparatus) is surveillance - «control of information and the superintendence of the activities of some groups by others» (p. 2). This process is much facilitated by a territoriality strategy, to use Sack's terminology, hence Giddens's definition of the state as a power-container (p. 172) :

> The nation-state is a power-container whose administrative purview corresponds exactly to its territorial delimitation.

The nature of the territorial container is thus central to the definition of the state itself, hence the importance of boundary disputes and cross-boundary incursions to modern nation-states: as Giddens expresses it (p. 291), they are «challenges to its administrative and cultural integrity».

Although territoriality is central to the concept of surveillance in his work, and it is thus an example of Sack's theory (Giddens, 1985, does not refer to Sack's 1983 essay, and Sack, 1986, does not refer to Giddens's 1985 book), Giddens does not directly address most of the central issues of the importance of territoriality to the nature of the state. This is the focus of an important essay by Mann (1984), however, which provides the clearest statement of the necessity of territoriality not only for the definition and operation of the nation-state but also for the autonomy that it has in contemporary capitalist society.

The state exists, according to Mann, to perform four tasks that are necessary either to society as a whole or to interest groups within it: to maintain internal order; to sustain military defence and aggression; to maintain communication infrastructures; and to achieve economic redistribution. These are most efficiently performed by the personnel of a state apparatus, he contends (p. 197), whose superior efficiency is a consequence of its territorial definition. The state, he argues, differs from other organizations in which power is exercised because only it «is inherently centralised over a delimited territory over which it has authoritative power» (p. 198). Thus :

> Unlike economic, ideological or military groups in civil society, the state elite's resources radiate authoritatively outwards from a centre

but stop at defined territorial boundaries. The state is, indeed, a place - both a central place and a unified territorial reach.

In its regulatory activities, it operates what Mann terms «infrastructural power», which refers to its ability to penetrate everyday life within civil society and implement political decisions (p. 189). Territoriality is a crucial strategy in this: the greater the infrastructural powers of the state, he argues (i.e. the greater its needs to penetrate everyday life), «the greater the territorializing of social life» (p. 208).

Territoriality is not only a necessary strategy in the operation of modern nation-state apparatus, however; according to Mann it is also the basis of its relative autonomy within capitalist societies, and hence its ability to act independently of various interest groups within those societies. Once the territorially-centralised state is established, he argues, civil society loses control over it; the state becomes an autonomous force, and its territorial definition is the basis of that autonomy. As Mann puts it, the achievement of territorial sovereignty by the state, with which is associated the monopoly of legalised violence, as identified also by Giddens, gives those in control of the state apparatus an advantage over others, some of whom may have access to greater stores of economic and other power. The latter may endorse the use of that state power, because they benefit from its exercise in general, if not on every occasion, and thus are prepared to sustain its legitimacy: where they do not, then there may be a challenge to the state, as illustrated by Habermas's (1976) work on crises in capitalist societies. Whilst society in general is prepared to legitimate the state because of the benefits that it brings, then the state elite are empowered to act autonomously, to do things that are not necessarily desired by those who still support it.

3 The role of the state and the uses of territoriality

The works of Giddens and Mann add to Sack's by providing an argument which poses but does not address relating to the necessity of territoriality in certain circumstances. With regard to the nation-state in capitalism, territoriality is a necessary element of its definition and operation, and a basis for the autonomous power that the elite in charge of the state apparatus exercise.

But why is the nation-state necessary, thereby making a territorially-based institution necessary to such societies ? Many authors have addressed this question, concluding that the continued survival of a capitalist mode of production requires an autonomous institution such as the state. Drawing on the work of O'Connor (1973), Clark and Dear (1984) have identified three basic functions of the state. First, it is required to secure social consensus, whereby all groups within the society agree (implicitly if not explicitly) to accept the way in which it is operated (what they term «the prevailing contract»: p. 43). This is its primary function, since unless the state provides order, stability and security, it is impossible for «production and exchange [to] take place with any degree of continuity». This is achieved, they contend, by three sub-apparatus within the state: the political sub-apparatus provides an organisational framework, via liberal democratic forms of government at national and local levels, for wide participation in, and therefore explicit acceptance of, the state's exercise of power; the legal sub-apparatus mediates in disputes and allows all to achieve their state-defined rights; and the repressive sub-apparatus (the police and the military) limits opposition and undertakes surveillance.

The second role is to secure the conditions of production, by investment that will increase production in both public and private sectors of the economy and by regulating patterns of consumption so as to ensure the reproduction of the labour force. In this, according to Clark and Dear (p. 43), the state is involved in providing «the infrastructure for economic growth and coordinated market exchange», and by doing that it «provides the conditions for creating profit, and hence ensures the allegiance of the capitalist elite» which «reinforces its own power and legitimacy». This involves sub-apparatus which produce public goods, contract with private sector organisations for the provision of other goods and services, and regulate money through the treasury; together these create an infrastructure within profits can be made and capitalism can flourish, an infrastructure which individual capitalist organisations could not provide, either separately or together.

The final function involves securing social integration, by ensuring the welfare of all and especially the subordinate groups within society who are economically relatively powerless. The second function involves the creation of wealth which is unequally distributed; the third involves the redistribution of some of that wealth towards the relatively underprivileged in society, so as to sustain their support for an economic system in which they do less well than others. This legitimation is achieved, Clark and Dear argue, by five sub-apparatus : those that deliver health, welfare and education services, those that distribute information, and those which facilitate communication.

Whereas a major task of the sub-apparatus involved in the legitimation of an unequal economic system is the distribution of a range of services, thereby winning support through the provision of material gains, they are also involved in what is termed «some degree of regulation or control over the serviced population» (Clark and Dear, 1984: 52). This is an ideological task :

> Ideologies tell people what exists, what is possible, and what is right or wrong. They structure the limits of discourse in society, and are present in all aspects of everyday life, including family, school, neighborhood and workplace... The integrative purpose of ideology is therefore to promulgate the belief that the «system» is capable of overcoming the contradictions of capitalist relations.

This may require a separate sub-apparatus, or it may be achieved by the others, as in the role of a state education system in the promotion of an ideology, for example, and the mass media in the sustaining of a culture which lauds the capitalist system.

This detailed analysis by Clark and Dear of the structure and functioning of the state in capitalist societies is not accompanied by an examination of the use of territoriality as a strategy by which its roles are (necessarily, according to Mann) exercised. The only discussion of territory that they provide is concerned with the local state, which is part of the apparatus of the state organisation (as examined in greater detail in Johnston, 1990b). They pose the question «why should it be necessary to create a small-scale spatial analogue of the national state ?» (p. 133), and answer it by asserting that it is because of the requirement for the state to be involved in «long-term crisis avoidance at the local level» (p. 133) in ways that «effectively co-opt and thereby control the local population» (p. 134). They do not address that need in detail, however, nor do they inquire why it is a necessarily spatial means of control. Their main interest is in the relative autonomy of the local state, and in establishing that the local government/administration systems are simply part of the state apparatus and in no way independent of it. By decentralizing at least part of the state structure and encouraging democratic participation in its activities, the central state elite wins the consent of the population and furthers the legitimacy of the entire state apparatus.

4 The role of territoriality in state activity

The various authors reviewed here provide elements of an approach to understanding the state through the concept of territoriality, but none brings the various parts together in an holistic argument that addresses the necessity of territoriality for an institution that is itself necessary for the long-term survival of capitalism. Sack has argued that territoriality is a sound strategy for the pursuit of control, which is central to the state's operations, but others have not incorporated that case into their work on the state itself.

Territoriality and the three state functions

Territoriality is a viable, probably necessary, strategy for pursuit of each of the three state functions identified by Clark and Dear. The justification for that assertion which follows here looks at each of the functions outlined above in turn.

Concerning the securing of social consensus, without order, stability and security, capitalism cannot flourish. People will not invest capital if the probability of making profits is low, because their assets may be confiscated, for example, or their workforce may be ill-disciplined and its productivity low. Thus, for example, certain countries may be much more attractive to outside investors and providers of development aid than others, with possible consequences for the nature of the state apparatus and in particular the operation of democracy

(Johnston, 1984, 1989a). The state must provide a secure environment within which wealth accumulation can be successfully prosecuted.

That secure environment must be one in which internal disorder is at a minimum and external threats are few, if any. The state must be able to deliver security and stability through «the rule of law», in which people either accept and operate within the existing laws or are coerced into so doing, usually with some promise that this will produce economic, social and political benefits in the future. In other words, the state apparatus must operate a successful surveillance function, which requires an efficient and effective police force: as Giddens argues, such an operation calls for a territorial strategy with the rule of the state's law prevailing throughout its defined sovereign area; that territorial strategy will undoubtedly involve a hierarchy of areas within which control is exercised, as made clear in Sack's examples. Similarly with external threats, the state must be able to defend itself against potential adversaries; its borders must be secure, which again requires a territorial strategy. (To some, the military strength of the state should be apparent but preferably not demonstrated. That a state apparatus is ready, willing and able to defend its territory should be sufficient to deter any would-be invaders. If it is called upon to demonstrate its preparedness, the costs of waging war may be a deterrent to investors because of the high taxation rates that may prevail; the decline of major political powers on the world scene is associated with some to the costs of waging expensive wars : Kennedy, 1988; Taylor, 1989).

With regard to securing the conditions of production, the discussion earlier identified the provision of an infrastructure as the central task. That infrastructure will have two components, a physical one - comprising, for example, a transport network - and a facilitative milieu. The latter is the more important, according to political theorists of the «New Right» who dominated the capitalist world during the 1980s. Their particular concern was with the role of the state as a guarantor of the operation of «free markets», to be achieved in a variety of ways among which one of the most crucial was the control of the money supply (Gamble, 1988). Inflation was identified as the scourge of capitalist economies then considered to be in crisis, and the proper role of the state apparatus was to remove it, thereby providing the conditions for substantial investment in the means of future wealth accumulation. If those conditions prevail, not only will wealth accumulation processes flourish but, according to some, private investors will provide parts of the physical infrastructure leaving the state's role as primarily that of creating and sustaining a facilitative milieu.

Arguments for the state as a regulatory body creating a favourable environment within which capitalism can operate do not necessarily call for a mosaic of nation-states such as that currently in place in the world. One state regulating the whole world would suffice. However, capitalism is built on competition between investors in different parts of the world, and while they all want a stable milieu in which to operate they also prefer one in which their own interests are protected somewhat. Hence, as Harvey (1985) argues, groups in particular areas will probably cooperate in what he terms a «regional alliance» to promote their interests or those in other places. (These alliances may combine either or both of the local capitalist and working classes, according to the particular conditions and how they are perceived). The pattern of nation-states inherited from the pre-capitalist era provides not only a useful set of defined areas within which those alliances can be formed but also a set of institutions which because of their relative autonomy, itself a function of their territorial identification, are able to undertake the regulation and to sustain it through their legitimacy. Thus the existence of a mosaic of states, allows the regulation of local populations in a whole variety of ways. The elements of that mosaic may not always be well suited to the task, as illustrated by the growing importance of the European Community in the regulation of the economic, and hence social and political, affairs of twelve formerly independent states, the successor to many previous nation-state-building efforts such as the creations of Germany and Italy in the nineteenth century

The state is not just an institution for the promotion of capitalist interests, therefore, but rather one which promotes the interests of those domiciled within its territory. It regulates the activities of individuals and corporate bodies through its sovereign control over that territory, using strategies as outlined by Sack. The same is true for its work in the securing of social integration : the redistribution of parts of the wealth created as a consequence of its activities under the previous function is confined to the residents of its territory : only residents of a state will normally be allowed access to the subsidised (even free) education and health services provided by a state, for example.

As discussed above, a major feature of this third state function is ideological in content, promoting a set of beliefs within the population which is supportive of the operations of capitalism in general and of local interest in particular. (The same has been true, and perhaps even more so, of the socialist states - dubbed by some as the welfare-state dictatorships - of eastern Europe between c1950 and c1990.) Thus in the United Kingdom in the 1980s, for example, there was a major government-led campaign to promote a pro-capitalist ideology, favouring the advantages of entrepreneurship and individual self-responsibility, through all levels of the educational system, which involved Ministers denigrating the views of many in the academic world. That campaign was more than pro-capitalist, however; it lauded local capitalist activity, sustaining the regional alliance of capitalist and some working class interests which supported the elected governments.

Nationalism as Territoriality

The promotional campaigns introduced in the previous paragraph are an example of nationalism, defined by Smith (1986b: 312) as :

> (a) A feeling of belonging to the nation; [and] (b) a corresponding political ideology which holds that the territorial and national unit should be allowed to coexist in an autonomously congruous relationship.

Johnston, Knight and Kofman (1988), drawing on the work of Anderson (1986), suggest that there are two types of nationalist campaigns, which differ in the relationship between territory and nation identified in Smith's definition.

In the first type, the territory of a nation and the territory of one or more states are not congruent, and the goal of the nationalist movement is to achieve congruence by a reordering of territorial boundaries : their strategy is clearly one of territoriality though, as will be elaborated later, it is not necessarily one of control over individuals. The nation is defined in cultural terms, and its representatives lay claim to a territory in which the nation will be autonomous, if not independent. The territory is occupied by others, and the goal is to remove them, as in current struggles in many parts of the world.

The second type of nationalism refers to situations in which there is no incongruity between national area and state territory, but the link between the two is weak. A territory over which a state has sovereign power is defined, but the residents of that territory identify only weakly with it; there is no enduring national sentiment. Thus the goal, usually of the elite controlling the state apparatus, is to generate and sustain national sentiment among the territory's residents, in part to legitimate the state itself and in part to promote the interest of the regional alliance(s) that it represents. This has been the case in many newly-independent ex-colonial state, for example, where the territories defined by the colonial powers bore little resemblance to the often weakly-defined pattern of national territories preceding colonial invasion; by the time of independence the possibility of returning to those earlier boundaries was remote, and so new national identities had to be built based on the colonial territories, a process usually initiated by the pro-independence movements that led to the creation of the new states.

The nationalism in countries like the United Kingdom during the 1980s referred to above is a sub-category of that described in the previous paragraph. The sense of nationhood among a substantial proportion of the population was considered insufficiently strong by those in control of the state apparatus for its program of economic, social and political restructuring designed to promote the interests of capital in general and the local regional alliance(s) in particular. A territoriality strategy was thus initiated to enhance people's sense of identification with the national territory.

Nationalism may be used as a strategy by which some people can control others, winning consensus support for a program from which there will be unequal benefits. In this sense it is a clear example of territoriality as defined by Sack. But nationalism is also an example of the need for identity, recognised by Smith in his definition of territoriality. If it is the case that people need to be able to identify with groups in order to define themselves (and social psychologists have shown that group identity is important to people's perceptions of the world), a territoriality strategy may be an effective way of providing a focus for self-identification. It is almost certainly not a necessary strategy, for many people successfully identify themselves with groups that have no territorial associations, such as religions (although, as Sack shows, most of these are territorially structured in their organisation, with territorially defined locales being used to mould people into communities with common interests : Sack, 1986: 87). In part it depends on the nature of the group and its goals. If it exists merely to provide a social focus to parts of individuals' lives, and if it is neither antagonistic towards nor antagonised by other groups in the same territory, then territoriality nay be an unnecessary strategy. But if there is antagonism, either one-way or mutual, territoriality may become a necessary strategy, as was the case in the cities of Northern Ireland after the onset of creased inter-religious community tensions in 1969. Further, if a group wishes some degree of autonomy or self-determination (Knight, 1988), territoriality may again be a necessary strategy; for groups lacking a territorial base, therefore, their self-determination ambitions are likely to be thwarted, which is the situation for many indigenous peoples throughout the world.

5 Territoriality and the definition of self and others

The examples in the last part of the previous section illustrate a use of territoriality as a strategy which extends Sack's theory in a way not explicitly envisaged by him. As he presents it, territoriality is a strategy by which people are controlled through inclusion, via surveillance mechanisms. By incorporating people within a territory you make them subject to the rules that apply there and so control their behaviour.

The creation of separate residential areas for different religious groups in Belfast (and, by extension, the creation of many other segregated communities on a variety of criteria) involves social control by exclusion rather than inclusion. The social relationships between the in-group and the out-group are controlled by restricting, and in some cases preventing, contact between them. Such a strategy may be operated without, the use of any of the state apparatus, through «voluntary» procedures by which people agree not to let members of certain groups into an area, but, as illustrated by racial segregation processes in American cities, these are difficult to police in certain circumstances (see Johnston, 1984). Use of the state apparatus can sustain such strategies, however, by, for example, land-use zoning practices allied to the incorporation of independent municipal entities (i.e. units of the local state apparatus).

The use of the state apparatus to promote residential congregation and segregation not only assists in the operation of social control via territoriality but also promotes the loner-term processes of unequal social reproduction in two ways. First, it facilitates the uneven provision of public goods, to the benefit of some and the detriment of others, as with the reproduction of social class differences through locally-provided education services in American suburbia (Johnston, 1984). Secondly, because separation restricts contact, it assists the development of inter-group stereotypes based on ignorance and fear rather than contact and knowledge (Johnston, 1989b). Those stereotypes promote images of a divided society, in which to identify oneself positively with the people in one territory is complemented by negative identifications of others, who live in other territories. (This operates at the national scale, too. The creation of a positive image of one's own nation is frequently associated with a negative image of others, leading to the mutual mistrust that underlies many international relations.)

Where territoriality is used as an exclusionary rather than as an inclusionary strategy for social control, therefore, the result is likely to be heightened social tension, hence the arguments for its abandonment (e.g. Sennett, 1970). One almost certainly leads to the other, however, for the creation of states using an inclusionary territorial strategy by its very nature involves the definition of «others». Thus the strategies identified by Sack as efficient and effective means for social control are also likely producers of social tension and conflict. They are, at a variety of spatial scales, the bases for geopolitics.

6 Forward

This essay has a beginning but no end; it has taken an important geographical concept and extended its interpretation, but has reached no firm conclusions. During the 1960s and 1970s, geographical conceptions of space focused almost entirely on it as a continuous variable (Sack, 1980), with the consequence that territoriality, concerned largely with discontinuous, bounded space, was largely ignored. Increasingly that is seen to be an error, and the growing interest in places at all scales (as reviewed by Gilbert, 1988, and Pudup, 1988) is correcting the balance. In that restructuring of the orientation of geographical work, territoriality is a central concept, and the discussion here has sought to advance its use by recognising as-yet little explored aspects of its relevance to the understanding on human behaviour.

References

Alford, R. and Friedland, R., 1985, *Powers of theory*. Cambridge: Cambridge University Press
Anderson, J., 1986, 'Nationalism and geography', *in* Anderson, J., ed., *The rise of the modern state*, Brighton: Harvester Press, 115-142
Ardrey, R., 1969, *The territorial imperative*, London: Fontana
Clark, G. L. and Dear, M.J., 1984, *State apparatus* Boston: Allen and Unwin
Dunleavy, P. and O'Leary, B., 1987, *Theories of the state*, Macmillan, London
Gamble, A. M., 1988, *The free economy and the strong state*, Macmillan, London
Giddens, A., 1985, *The nation-state and violence*, Cambridge: Polity Press
Gilbert, A., 1988, 'The new regional geography in English - and French-speaking countries,. *Progress in Human Geography*, 12, 208-228
Habermas, J., 1976, *Legitimation crisis*, London: Heinnemann
Harvey, D., 1985, 'The geopolitics of capitalism', *in* Gregory, D. and Urry, J., eds., *Social relations and spatial structures*, London: Macmillan, 128-163
Johnston, R. J., 1984, *Residential segregation, the state and constitutional conflict in American suburbia*, London: Academic Press

Johnston, R. J., 1989a, 'The individual in the world-economy'. *in* Johnston, R. J. and Taylor,
    P. J., ed., *A world in crisis ?*, (second edition), Oxford: Basil Blackwell, 200-228
Johnston, R. J., 1989b: 'People and places in the behavioural environment', *in* Boal, F. W. and
    Livingstone, D. N., ed., *The behavioural environnement*, London: Routledge, 235-252
Johnston, R. J., 1990a, 'The territoriality of law: an exploration'. *Urban Geography*, 11
Johnston, R. J., 1990b, 'Local state, local government and local administration', *in* Simmie, J.
    and King R.,eds., *The state in action*, London: Belhaven Press
Johnston, R. J., Knight, D. B. and Kofman, E., 1988, 'Nationalism, self-determination and the
    world political map: an introduction'. *in* Johnston, R. J., Knight, D. B. and Kofman,
    E., eds., *Nationalism, self-determination and political geography*, London: Croom
    Helm, 1-17
Kennedy, P., 1988, *The rise and fall of great nations*, London: Fontana
King, R., 1986, *The state in modern society*, London: Macmillan
Knight, D.B., 1988, 'Self-determination for indigenous peoples : the context for change', *in*
    Johnston, R. J., Knight, D. B. and Kofman, E., eds., *Nationalism, self-determination
    and political geography*, London: Crom Helm, 117-134
Mann, M., 1984, 'The autonomous power of the state : its origins, mechanisms and results',
    European Journal of Sociology, 25, 185-213
O'Connor, J., 1973, *The fiscal crisis of the state,* New York: St Martins Press
Pudup, M. B., 1988, 'Arguments within regional geography', *Progress in Human Geography*,
    12 369-390
Sack, R. D., 1980, *Conceptions of space in social thought,* London: Macmillan,
Sack, R. D., 1983, 'Human territoriality; a theory', *Annals of the Association of American
    Geographers*, 73, 55-74
Sack, R. D., 1986, *Human territoriality: Its theroy and history*, Cambridge: Cambrigde
    University Press
Sennett, R., 1970, *The uses of disorder*, London: Penguin
Smith, G. E., 1986a, 'Territoriality', *in* Johnston, R. J., Gregory, D. and Smith, D. M., eds., *The
    dictionary of human geography*, (second edition), Oxford: Basil Blackwell, 482-483
Smith, G. E., 1986b, 'Nationalism', *in* Johnston, R. J., Gregory, D. and Smith, D. M., eds., *The
    dictionary of human geography*, (second edition), Oxford: Basil Blackwell, 312-314
Taylor, P.J., 1989, *Political geography*, (second edition), London: Longman

Ron J. Johnston
University of Essex
Wivenhoe Park
Clochester CO4 3SQ
United Kingdom

15 The spatial and the political : close encounters

Jacques Lévy

The spatial and the political have been particularly affected by the turbulence and progress that social sciences have been undergoing. They have been the subject of rethinking from the inside and discussion from the outside. It is now important to turn these innovations to profit in more concrete terms. With this end in view, we shall put forward a few remarks on the relationship between the spatial and the political, as being two dimensions of the social totality. The linking of the two will not be envisaged in the mode of an intersection (which would belong to each of both sets at once), but in that of perspective. If it is accepted that society cannot be cut up into slices without changing its taste, then a good cooking method consists in combining the ingredients in their entirety according to an original recipe. At this point we shall ask ourselves about the advantages there might be in providing the geographical approach with a political «filter» (in the photographic sense) and in observing society geographically while enhancing its political «colour». And we shall attempt to estimate what part of social reality has thus been captured, treated and fixed.

1 From factor to system : the spatial and the political within society

Looking at the political from a geographical angle implies having a fairly clear idea of the limits of each domain. This is not an academic question: the polysemy around these terms is such that, from glide to glide, it would be easy to fit everything into them. We would find ourselves in the situation of the Sixties where separate voices proclaimed that everything was political and that nothing escaped geography.

227

G. B. Benko and U. Strohmayer (eds.), Geography, History and Social Sciences, 227–241.

Space, from preconception to concept

To say that space is a category corresponding , according to Marxist tradition, to *a mode of existence of matter* (axiom A1), is to contest from the outset that spatiality can be defined in itself, independently from the «contents» of reality that it organizes. What had long been the founding postulate of geography, in its Kantian form (space *a priori* giving its form to things, which suited «natural determinism» quite well), then its Cartesian form (geometric laws for an immanent space, claimed by followers of spatialist modernism), have gradually given way to a new notion. The immediate consequence of this axiom which few would now dispute, is that the major categories in the spheres of knowledge are going to lead to a *partition* in the types of space : physical, biological and social; understanding space in societies can only be done, unlike for other aspects of their existence, by means of a social science.

Then the problem of the relevance of a geographical approach presents itself, without *a priori*. Simple deductive reasoning is not enough to convince us that the spatial *field* can become an *object*. Yet social space, that is to say the distribution of social phenomena over the two curved dimensions of the earth's surface, does not necessarily confront society with an interesting problem. If localizations are made practically imperative by the lack of human control over living conditions , we are faced with a *pre-spatial* society; if, on the contrary, the freedom to allocate elements of the society here or there is total, if material ubiquity (transport) or immaterial ubiquity (telecommunications) generates a total isotropy, we can say that this society is of a *post-spatial* nature. Thus it can be set forward that it is in the transitory phase between these two extreme situations that a scientific approach to social space becomes meaningful (Axiom A2).

Once this second axiom has been accepted, the problem presents itself of how the different forms of knowledge about society interlock. After a long period of acceptance dating back to Pareto or even the physiocrats, the organizing of causality into a hierarchy with emphasis on the economy is widely challenged today, even by those who take their inspiration from Marxism (Godelier, 1984). On one hand, the level of development of a society, its «productive forces» which make possible all forms of production, including that of the society itself, cannot be reduced to an easily measurable, and certainly not monetary, «exchange value». On the other hand, it appears that none of the usual ways of understanding society - the economical, the sociological and the political- can be conceived of without the *exteriority* represented by the other two. Trade relations cannot be understood without reference to social relations, nor can the former be understood without recognizing the existence of a political power. There cannot therefore be a hierarchy between social sciences since each of them is capable of taking into account the whole of social reality and each of the phenomena that constitute society (Axiom 3).

(Axiom 4) For, it is not possible to isolate a group of «things» from society in such a way as to understand their workings without bringing the rest of society into our argument. All social sciences are thus both total and partial; they represent a dimension which is also a problem for society - freedom and constraints, choice and stakes -which provokes reflection in some of its members.

Thus 4 axioms indicate the progression which leads space from preconception to concept, while identifying it as a philosophical *category* , a historical *field*, one *dimension* of a plural knowledge of the social, and a discourse on society as a *totality*.

The political: from things to the relationship between things

If the last two axioms apply to the political dimension, the scope of the latter remains to be defined while taking into account both the present state of knowledge in the matter, and the diversity of the study material available from societies which exist, have existed or could exist.

By emerging gradually from a «political philosophy» which was more moral than scientific, we have learnt to distinguish *political* phenomena from their material or almost material manifestations, such as the state. J. Benda (1965) has already given a wider vision by talking about the political , while anthropologists (Balandier, 1967) have explored politics without the state. Conversely (Cot, Mounier, 1974), it does not seem useful to assimilate all «power» - over things or people - to political power. The definition of the political has thus been reorganized around a new centre , becoming more complex and more immaterial while distancing itself from an individualist vision controlled by «appetites», as Machiavelli saw them, but also from too rigid an association between capacity for using physical force and political power- which had been put forward by Max Weber or Lenin. As Gramsci had realized with the example of his own society and as others have confirmed since, it appears indeed that, even in the most brutal dictatorship, the exercise of power consists above all in managing consent and dissent, in gaining or keeping a minimal degree of hegemony (in Gramsci's meaning of the word, that is, ultimately a phenomenon situated in the realm of representation) without which even the deadliest arms have little weight. The political emerges therefore as a function of overall control arbitrating at society level between the other , economic and social functions. The political begins where there is *societal legitimacy* , be it real or potential, where there exists a claim to organize the divisions of society in such a way that, in a given system of finalities, its unity is reinforced.

The political can still less be reduced to politics, which limits itself to the offer of political discourse within the state (Bourdieu, 1980); the *market* as a whole has to be considered (offer + demand + relations between these), but also what comes beforehand : the «modes of production» this offer and demand, and finally the tension which exists between this market and the actual practice of political power as well as , beyond that, the overall dynamics of the society, in short the whole relationship between Gramsci's «civil society» and «political society». In accordance with G. Bergeron (1977), who distinguished between *politics*, the political, polity, «governance» and «regime», we would do well not to confuse the political with policies, which are certainly an element in approaching the political but which partake of other forms of logic, once international politics are involved. The action of one state in relation to other states presupposes that its own internal legitimacy be achieved, even if the former contributes to strengthening the latter. As there only exist fragments of a world political system (of which «international political opinion» is a sign) in which the destiny of dissent and consent would be played out on a world scale, a wide area subsists where only domination, control and influence play out their logic : this is the domain of international strategy, with, in the spatial dimension, geopolitics, where originally military concepts can be applied. The same is not true of the interior of a political system - unless it is considered that military victory implies the prior consent of the defeated. The political definitely cannot be confused with the problems of the State.

On the other hand, the political opens out on to sociology, for the relation to politics is a social relation. It contributes, as far as groups are concerned to actions of positioning and evolution in society, thanks to its weight in the overall regulatory function. «Demand» in politics, studied by political sociology, refers to representations and practices that characterize the identity of these groups : their positioning within the network of groups, their vision of themselves and of the world, and their strategy. On the contrary, the explanation for modes of behaviour has long been sought by political scientists within the logical systems of institutions- those of power games as well as those of the means of delegating power. We can appreciate how interesting it can be to «cross» over, in one movement of thought, the systems of choice and risk, and to understand this interface otherwise than in terms of a «black box». The spatial offers this transversal movement.

Towards a dialectic systemism

The first temptation which awaits someone who wishes to envisage the political from a spatial point of view is to add on space as supplementary «factor» in explanation. On the subject of the exercise of power, administrative or electoral divisions would thus bring their complexity into account in order to qualify one or other overall appraisal. Even more obviously, in political behaviour, space would add to the already well-established contingent of indicators having already proved their worth : socio-professional categories, religious beliefs, sex and age, or those more recently focused on, like patrimony or approval of certain types of values.

It is true that the legacy of A. Siegfried for a long time (and lazily) used as a simple search for «permanents» (Goguel, 1970; Leleu, 1971), opens out on to the rediscovery of the spatial approach to political phenomena (Guérin, 1984). Spatial analysis (Groupe Dupont, 1982; Rapetti, 1985) constitutes above all a challenge to over-simple explanatory models. In Paris, for example (Lévy, 1985), the study of votes by «arrondissement» shows that we cannot merely accept the notion of «gentrification» as sufficient explanation for the change in political behaviour.

However, we must go further than the accumulation of «factors», each supposed to explain a supplementary part of reality, and this , independently of the qualitative or quantitative techniques of data analysis. It is ,first of all, necessary to take care in the use of statistical notions : averages, medians, standard deviations, all of which are valuable if one is capable of giving an internal meaning to the object under study - which is never self-evident. After that two levels should be clearly distinguished : the *empirical* where *variables* are used to cut up reality into objects which are close to those of common understanding and well-represented in statistical sources; then the *theoretical* level of *conceptual determinants* which, when combined, make up an explanatory model. Without taking up , as Popper and his successors have, the classic discussion about the difficulties of experimental verification or, as Bachelard, Canguilhem, Koyré or Althusser have, the debate on the specificity of scientific reasoning, we will content ourselves with noting that, if we confuse empirical values with causes, we end up, after apparent initial successes, with insurmountable difficulties : the inability to conceive of, or even less foresee changes in the object under study; the incapacity to define areas for implementation of *action* on this object, especially if is ambitious. Thus, the knowledge that there exists a strong link between voting for the Right and being a practicing Catholic does not permit us to understand either why practicing Catholics vote for the Right or why at certain times (as in the west of France in the Seventies and Eighties) a section of them began to vote for the Left, or still less, how to encourage or prevent this trend. To go a step further, we must try to build up the individual and collective logic systems and the world-view of protagonists for whom voting and religious practice may only be by-products of principles which may or may not be coherent with one another, but which are pre-existing (Michelat, Simon, 1977; Michelat, Simon, 1985).

Advancing along the explanatory path brings one to adopt a *dimensional approach*. Space is not a supplementary factor, but a means of interpreting the totality of society. It consists in asking oneself up to what point the physical distribution of society's institutions allows one to perceive their functional and developmental logic. If it is taken for granted that , in a social structure, the determinants organize themselves in contradictory relationships of interdependence, this must also be true for a spatial approach to them. As for any scientific construction of society, we are forced to take into account, in each problem, however small, its being part of a whole. In this sense, such an interpretation is necessarily systemic. This term in itself would justify much debate. By speaking of *dialectic systemism*, we will merely show our refusal, at the same time of mathematical or literary empiricism and of functionalism, and we will put forward that society is a totality ridden with contradictions and whose very movement is made up of their causes and effects. In the political sphere, this leads notably to the consequence that in the systemic approach we must beware of «discovering» too easily a «harmonious» working of the social order. Faced with the theoretical naiveté of people like Merton, Easton or Almond, a preoccupation with contradiction should help us to avoid ethnocentrism (Bourdieu, 1972; Gaxie, 1978).

Making claims for the existence of forms of *spatial logic of the political*, liable to explain why one votes or organizes differently here and there, is not suggesting the extra-social treatment of social phenomena; it is constructing chains of causality in the political by taking localizations and their reciprocal relations as a mode of organisation of the social. In this way we define a potential field which should be validated as regards the axiom of relevance (A2). What can space have to say about the political ?

2 Spatial logic in the political

The idea can be upheld that spatiality is built around a single notion, that of distance (Gatrell, 1983). It is from the coexistence (according to Leibniz's term) of two phenomena in the same place, or, on the contrary, in two distant places, that arises the differentiation of the diverse points of a given area. This simple reality (without distance, there is no space) must be recalled whenever we question the relevance of a spatial relation.

Distance and link

There exist four simple types of possible spatial relations between two social phenomena. By phenomenon we mean here not an empirical indicator, but an active element (protagonist, institution, process),separated from the totality of social forces in order to be presented as an explanatory account.

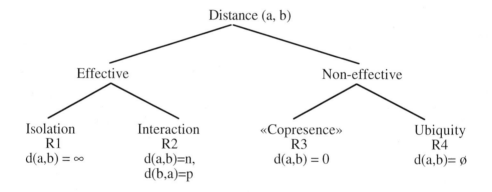

Figure 15.1 : The four spatial relations

These are the basic structures which, when combined, produce more complex configurations. Thus, what we call interaction (R2) describes every possible effect of one phenomenon on another situated in a different place. This influence may result in the extension of the phenomenon a or different modes of integration which have been analysed by A. Reynaud (1981). The uniting of the various relations, in particular R2, creates spatial systems or networks, which can in their turn be studied as such (Durand-Dastès, 1984). As for the R1 (pre-spatial) and R4 (post-spatial) situations, these are only valid in a spatial approach in so far as they are confronted with the others; then they represent exceptional cases from the point of view of relations to distance.

Using this simplified diagram, we can ask ourselves if spatial relations affect - and if so, in what way - the political processes.

From ubiquity to interaction

We could follow the history of the internal administrative divisions within states according to the means employed by their heads to *control* the most peripheral areas. We would show how, from the initial isolated little groups, we move on to more decentralised (feudal) systems or less decentralised («Asian») societies, ending up, thanks to the use of sufficient resources, with the effective ubiquity of the State. Today, it seems quite clear that the notion of control does not exhaust the complexity in the processes of adjustment in internal politics.

The gradual and modulated arrival of «universal suffrage» alters the scene not only by creating a new field for political sociology, but also by modifying the meaning of the political institutions themselves. The analysis of the electoral manipulation obtained by gerrymandering (Taylor, Johnston, 1979) are certainly not without interest; but we could not reduce the spatiality of the relation between state and electors to this. Relevant divisions from this point of view are not necessarily a carbon copy of the apparent centres of the exercise of power. In France, a country until recently very centralised, the Prefect should have represented the sole effective infranational grade, since , until 1982, he had at his disposal all government powers at a local level. In actual fact, institutions with little financial and administrative autonomy and which were officially apolitical, the *«conseils généraux»* (Departmental authorities) and *«municipalités»* (town councils), had played a considerable role for a century because they had organised consent and dissent. Paradoxically, the main role of the Prefect consisted in «preparing the elections», that is to say, in understanding the local power struggle and political stakes in order to give central government the means of influencing them.

As soon as it becomes important to guarantee constantly a benevolent neutrality on the part of the majority of the electorate, reproduction of the exercise of power depends on successfully structuring overall societal aims and the institutions, whether large or small, which are the object of the different protagonists' strategies. Here institutions of power delegation within the state are obviously concerned, but also those *outside the state* which contribute at different levels to the shape at any one instant of regulatory constraints. The institutional space - if we are willing to look on it with a modern eye- organizes, in the same group of places, devolved and decentralised powers, political apparatus as well as ideological (the media, religions, the evolution of ideas), trade unions and other «legitimate representatives» of social groups, businesses and their heads, not forgetting the lower-level authorities. As J.-P. Guérin (1984) has shown, the development of winter sports in the French Alps, which appears to have resulted from a centralised political decision, can also be interpreted as the outcome of a consensus between the protagonists belonging to the different social groups concerned on the notion of «progress». In the same way, I have tried to show (Lévy, 1983) in the great shift in French urban planning in favour of houses rather than flats between 1960 and 1980, the effective decision-makers were as much the social groups, from the «top» to the «bottom» of the hierarchy, as the institutional operators who managed -rather than generated the demand.

We can understand the necessity of only considering as a *symptom* the discourse of politics on itself, which most often presents itself as the sole agent of social change, thus entering the field of discourse versus practice, and even that of manipulation (Taylor, 1984). The spatial analysis, as long as it does not forget any protagonists along the way, is usefully demanding. It is necessarily complicated, since every element of the system has its own spatiality : the elected body, the local newspaper, the employees of a big company, the society for the protection of a «threatened» area each define their space which combine with all the others to give an image of power in action.

Choice : identity and diffusion

If we now take the stand-point of demand, that is to say, of the different social groups' relationship with politics, we will quickly discover that we are confronted with comparable problems. Here again we witness , somewhat later than in the case of direct state control, the conquest of ubiquity. The opening up of remote areas, the growing spatial dimension of social mobility which itself is expanding, gives rise to a *de-spatialisation* of the sources of opinion-making. In the nineteen-forties, P. Lazarsfeld and his team had encountered without actually seeking it, what E. Katz (1957) has called t*he two-step flow of communication* : the flow issuing from the centre only reaches individuals through local notables who adapt it to local conditions. It would be invaluable to learn how this process has evolved in recent years. The narrowing of the maximal gap between the support in different areas for a given party can be explained by the disappearance of isolated groups: all the parties have become national, unless their existence depends on being confined to a certain space (regionalist parties). At the same time, however, we have to admit that the repetition in national elections of a divergence of 1 to 4 for a given party, or even of 1 to 7 for the small or extreme parties, between areas which are nevertheless quite vast (cf., for Western Europe, the study by J. Vanlaer (1984), means something other than a gradually acquired «legacy». This therefore confronts us with the problem of *resistance*, that is to say of the continual genesis of strong local or regional *identities* (an expression of the R3 relation) for which the usual socioprofessional classification is inadequate.

Thus in the case of Italy before the events of the early 1990's (Brusa, 1984), it has been possible to speak of subcultures which, within each spatial unit concerned, combine the different «ideal» and «material» elements of the relationship to politics as it is conceived and acted out by the citizens. We can then contrast the Christian-Democrat clientele (or sometimes neo-clientelist with the PSI) in the Mezzogiorno, based on the direct exchange of votes and favours, with the «secular» vote (PSI,PRI, PLI, PLI, PSDI), in the big urban areas of the Milan-Turin-Genoa triangle, which work more along the lines of ideological support, and with the *subcultura rossa* by which the PCI brings together in its bastions of Emilia-Romagna-Umbria a «vote of adherence» from the rural communities, a «vote of opinion» from the urban intellectuals and an excellent integration of numerous related groups among the unions, associations and management (Brusa, 1986).

Yet the dynamics of these once homogenous areas leads us to envisage the opposite phenomenon, in which forces situated in one point in space exert their influence over a surface of variable size (R2). For such a phenomenon to show itself, there must exist between point and surface a «difference in potential», a sufficient «slope» to counter the external noises which are liable to interfere in the relationship or even to cancel it (R4). Thus we can measure the relations between regional centres («*chef-lieux*») and their province ((Brusa, 1984) and ask ourselves about their spatial significance, while being wary of the apparently obvious. D. Rapetti (Rapetti, 1985) has thus been able to show that in Nantes, in 1969, the area of distribution of *L'Humanité* only matched in an imperfect fashion the map of Communist voting.

As a means of clarifying these situations, the Portuguese example represents a sort of working outline, thanks to the simplicity in the spatial structures of political behaviour (see map on facing page). In the case of Covilhã, a small region of textile industries embedded in the «deep» conservative North, M. E. Arroz and J. Ferrão (1979) have shown how the strong influence of the PCP locally had developed from one single source (*foco*), the working-classfreguesia (parish) of Tortosendo. On another scale, there seem to be strong relations between the southern suburbs of Lisbon and the countryside of the Alentejo, both of them overwhelmingly pro-Communist. Here the greater distances seem to be connected to the continuance of precapitalist structures in the means of reproduction of working forces: the town has remained to a certain extent dependent on the countryside, at least in its representatives. This type of diffusion would be reversed as soon as the break in community ties became effective and no longer prevented the extension of the relationship between the centre and the periphery. The «front lines» between the large «subcultural» groups can also provide detailed applications of the diffusion model. This is the case of Rio Maior (Burguette, 1978), a small «white» town where the riots started in the summer of 1975 after what the inhabitants considered as an intrusion (in the form of a public meeting) by the neighbouring small town of Alpiarça, vanguard of the «red» zone. Here we can speak of «negative diffusion», when the metaphor of territorial conquest «naturalizes» at a local level national political stakes.

The partisan interface

The parties present a panoramic observatory for exploring the political space. On the one hand, we are faced with pillars of the political system, instruments of the conquest and exercise, either direct or indirect, of the State's power. On the other, they constitute a means of refraction and thus of representation, of their electors, whether real or assumed, thanks to their concentric «rings of participation» (Duverger, 1981), from leaders down to voters. This representation is admittedly altered by the reciprocal determination in relation to one another (Bourdieu, 1981); it is peculiar because of the internal logic of each party and their strong oligarchic tendencies, as the classical writers (M. Weber, M. Ostrogorski, R. Michels) have amply demonstrated. However «total» the party institution (Verdès-Leroux, 1981) may be, it allows through, like an *airlock*, at varying rhythms and magnitudes, the tides emanating from the «higher» or «lower» reaches that make it up.

Thus the map of internal voting in the federal conferences (i.e. by *Département*) held in preparation for the 25th Congress of the French Communist Party, the PCF (in February 1985) which, for the first time, expressed the emergence of a substantial opposition to the outgoing national leadership, provides a complex illustration of the different party to society relationships : weak and therefore «aided» federations, dependent on the national leadership (R2; Vendée); federations developing solid relationship of identity with local society (R3; Meurthe-et-Moselle) and cutting themselves off in their isolated group (R1; Corsica) or arguing with the centre (R4) over their interactive influence on neighbouring areas (Haute-Vienne). With this last point we are confronted with the problem of «contagion», the phenomenon which makes an area resemble its neighbour more than if random distribution existed. This *neighbourhood effect* had been suggested by K. Cox (1969) at a local level; it has been amply re-examined by R. Johnston (1985) who has tried to understand by what actual processes relatively homogenous spatial groups are formed.

3 Scale and relevance

The word «scale» indicates both the *degree of magnitude* of an object and the relation between the metrical space of a map and that of the land represented. This polysemy in everyday language should not be sufficient to make one believe that phenomena are only given a size once they are mapped. Any scale (of phenomena) can be represented on any (cartographic) scale. The rest is a question of handiness, not of epistemology, since geography no longer claims to be the «science of the visual». Another ambiguity concerns the relations between the *manifestation* and *explanation* of a phenomenon. A spatial structure can be explained by means of causality whose territorial presence is much wider than that which we are seeking to explain; it may even be a matter of a process that has freed itself from the constraints of distance (R4). And, indeed, nowadays, it is always possible to show that a phenomenon undergoes , to a greater or lesser extent, the influence of other, larger phenomena, right up to the world level. But nothing is to prevent us from thinking that there exist *thresholds* beyond which we can obtain, without leaving the defined spatial framework, the highest quality of self-explanation of the social by the social. The national level certainly constitutes one of these thresholds.

In so far as the entire range of social functions are interdependent, the taking apart of a precise phenomenon is likely to give rise to a dividing up of space which will have significance for the whole of the society under consideration. There can be no question of defining *a priori* , and certainly not in terms of km or km², levels of scale which would fit mechanically into each other. On the contrary, we could speak of *relevant spatial levels* when the dividing up of space defines one or more distances, a specific metric theory of the series of phenomena and of the society under study. This is what we would like to reflect on regarding political space.

The local : the town or village knot

An apparently simple approach consists in moving up from the smallest spatial unity where «the political» manifests itself. This plan becomes complicated by the multiplicity of possible answers. Often the smallest institutionally competent area is the town or village, but the sizes vary from country to country, as within the same state. Countries with very widespread or populated towns or villages have sometimes created lower levels of authority (the *juntas de freguesia* in Portugal, the local area councils in Italian towns)with limited jurisdiction. It is not even necessary that this spatial unit be officially invested with power. The division into constituencies in the United States, in Great Britain or in France during the 3rd and the 5th Republics (except 1986) is enough to define areas seen as significant by the local protagonists. This is even more obvious in the pre-unified Federal Republic of Germany and in Italy where, because of the aggregate distribution of the proportional vote, a preferential vote for a particular candidate is «disinterested» from the point of view of national choices; his or her ability to become identified with an urban district or a small rural «region» can be played to the full and the emergence of the local becomes automatic (Brusa, 1980). We thus dispose of a great number of equally local divisions which nevertheless remain within the framework of the State. Other power centres must also be taken into account, the local pages of newspapers or the management of sports teams ...(Augustin, 1985).

From the political sociology angle, we can seek out homogenous units, which is not simple, since spatial segregation is rarely total. Marking the limits of a small urban district presupposes a reflection in terms of *sociability* (Frémont, Chevalier, Hérin, Renard, 1984). Taking into account the weakening of «transcendencies» (Barbel, 1984), especially of membership groups, it is in the *individuals* «group» that we must most probably look for one of the nuclei of sociopolitical spatiality. But it is then impossible to elude an unavoidable, if in many ways dissymetrical, *duality*: the local means at least living place + work-place. Yet each center of attraction has its own modes of political socialization. A variable as simple as the size of a firm allows us to gauge the amount of internal politicization. As for the living place, it is unnecessary to list all the politically significant applications, from housing to bowls. If both localizations coincide for a large section of the population, the two structures can mutually influence each other and create an original system such as those of the «red suburbs» of the cities of southern Europe, including France. If, on the contrary, a gap forms between these two systems of logic - and it is precisely here that the political stakes are to be found- the local political scene can no longer present itself as a scale model of the problems of the whole society; it has to imagine itself as a small cog in a machine against which it can only summon the strength of the weak : by encouraging solidarity or by filling as yet unfilled spaces.

Faced with this maze of divisions, we can adopt a deductive approach by projecting large geographical divisions on to the political space. Separating the country into *rings of urbanity* (Lévy, 1983; 1984; 1994) allows one to envisage Paris not only as a town in the administrative sense - which it has only recently become - but also as an overall space in opposition to the suburbs and to the countryside and compared with other centres in France and abroad. This central urban identity, which relies both on specific practices and on their portrayal, is a better explanation of the marked unity in behaviour of Parisian electors for the last quarter of a century than the socio-professional approach. In the same way, the transformation of space by the spread of urbanization is an opportunity to get down to the bare bones of local political life. P. Dressayre (1980) has shown, in relation to a small Breton town, how the changes in population lead not only to a renewal of representatives, but also to an alteration in the relationship with politics, even though the institutional changes are apparently minimal.

There emerges from these two French examples the *crystallizing power* of local problems at a town or village level. Problems that are below or beyond its explicit jurisdiction are nevertheless focused on the town or village, since this is seen by all the protagonists as an intermediary between the action taken by organizations and social groups and involvement in actual management. The strength of consensus gives rise to very efficient self-perpetuation, in spite of only slight technical justification for the intensely rural area or the large conurbations (replaced here by the *«Départements»*, districts and «Urban Communities») and in spite of repeated but ineffective attacks from central government. Even if the French case represents a sort of borderline in this area, it is possible to think of the town or village , in many other countries too, as constituting an important *societal knot* and probably the smallest relevant level of political space.

The regional: still sketchy

The intermediary levels between the local and the national present the difficulty of not always having strong, clearly-defined, stable institutions at their disposal. Some, it is true, appear quite clear-cut: the West German *Länder*, the Swiss cantons, the States of the USA, the provinces of Canada. These units are all the more able to mark out the national territory as they can avail themselves of substantial financial and political means; they are an echelon of the exercise of power as well as that of choice (Paddison, 1983). Sometimes the federal system allows the first infra-national level to fight with the central power for some of its most usual attributions : foreign trade or international sovereignty. This is because, as in the case of Canada, the national level has to gain a place in the sharing out of power that it had not acquired at the outset.

We can possibly consider as negligible the «noises» made by lower echelons (the «white» province of Lucca failing to give a «pinkish» tinge to Tuscany). We can also allow that there exists a superior division which corresponds to subcultures: the coherence of the Southern United States does not exclude that of South Carolina.

It is more difficult when the administrative regions do not exist, as in France in the 1950s and 60s or, as in present-day Britain, when they have effectively no political jurisdiction. In the absence of a functional interface between society and power, there is competition between several types of potential levels with possibly this intermediary level being inexistant. Thus R. Dulong (1978) contrasted the Limousin to the Lorraine: in the former, the designated partner of the local farming communities was directly the central State - government, economic structure and view of society; in the latter the preoccupation with *regional political autonomy* was emerging well before it was satisfied. This means that basically the modern region - a localized community of problems managed by those directly concerned - occurs within a historical evolution having a beginning, an end and numerous variants (Lévy, 1981), hence the Spanish *à la carte* approach to regionalization. It is no use trying at any price to find a region which would have no meaning.

By accepting that in developed countries of a certain size, the moment is right, because the regional level represents a good compromise between the high mobility of objects and agents and the far lower mobility of protagonists, between fast-moving socio-economic dynamics and a renewed fixing of identities, then the region stands out, with or without official institutions. This evolution can rely on different characteristics which are likely to unify areas which had up till then considered themselves as distinct: the defense of a threatened social group (Languedoc), the encouragement of a language which has long been repressed (Wales, Brittany), emphasizing particular local handicaps (Sardinia), and above all, demands based on a strong element of nationality (the Basque Country, Quebec, Belgium, Corsica, Scotland...). In each case the signifier is less important than the signified, the metaphor than the stakes: it is the acquisition for all or some of the local political protagonists of a greater share of regional autonomy than they have, possibly at the cost of going beyond common law (special status or independence).

The map of political *regionality* still reveals large blank spaces and noticeable gradients. While the local fans out from conquering individuals and stable institutions, the superior level or levels remain sketchy. In order for a regional division to build up a *permanent sub-system*, resistant to both the inferior and superior echelons, there must be both a «maturing» of society and a «helping hand» on the institutional side. There is no denying the usefulness of considering the relevant spatial levels, this one as much as the others, not as a simple framework for analysis but as a *spatial stake* in social evolution - and just as much as a *societal stake* in the dynamics of space.

«No simple truth exists, but it does not follow that truth is dull», wrote A. Grosser (1972) about political explanation. It is a way of saying, as G. Bachelard does (1971), that «science simplifies reality and complicates reason». With four axioms, four relations, a local knot, a regional «sketch» and a few hypothetical leads, we may believe that we have both simplified things and made them more complicated. When, at the beginning of the century, a form of political geography could have developed, the geographers of the time found it too theoretical, and to be honest, too, well... political (Guillorel, 1983); the political analysts developed one on their own lines, without any argument but without conviction. «Electoral geography», that old stylistic composition on the myth of «permanence», foundered in the face of accelerating social dynamics, whether seen from the spatial or political angle. At present, the epistemological horizon in social sciences has cleared, the theoretical foundations of space in society have been consolidated, the ideological obstacles and the empirical impasse are fading. And inasmuch as the protagonists of the spatial and the political have in future some reason to be wary of their institutions on their own, social demand is progressing. The understanding of the spatial structure of the political processes has become a credible project.

References

Arroz, M.E., Ferrao, J., 1979, 'A difusao do voto a partir de focos partidarios : o caso do conselho da Covilha', *As eleiçoes legislativas / Algumas perspectivas regionais*, Lisbon: Horizonte

Augustin, J.P., 1985, 'Pratique sportive et perception de l'espace', International round table, *Pratique et perception de l'espace*, Pau: Hegoa

Bachelard, G., 1971, *Le nouvel esprit scientifique*, Paris: PUF

Balandier, G., 1967, *Anthropologie politique,* Paris: PUF

Barel, Y., 1984, *La société du vide*, Paris: Seuil

Benda, J., 1965, *La trahison des clercs*, Paris: J.J. Pauvert

Bergeron, G., 1977, *La Gouverne politique*, Paris-La Haye: Mouton; Québec: Presse de l'Université Laval

Bourdieu, P.,1972, *Les doxosophes*, Paris, Minuit

Bourdieu, P., 1980, 'La représentation politique', *Actes de la recherche en sciences sociales*, 36-37

Brusa, C., 1980, 'Problemi di ricerca in geografia elettorale', *Lombardia nord-ouest*

Brusa, C., 1984, *Geografia elettorale nell'Italia del Dopoguerra*, Milan: Unicopli, 2nd ed.

Brusa, C., 1986, *Elezioni, territorio, societa*, Milan: Unicopli

Burguette, M., 1978, *O caso Rio Maior*, Lisbon: O Seculo

Cot, J.-P. and Mounier, J.-P., 1974, *Pour une sociologie politique*, 2 Volumes, Paris: Seuil

Cox, K., 1969, 'The voting decision in a spatial context', *Progress in Geography*

Dressayre, P., 1980, 'Suburbanisation et pouvoir local', *Revue française de science politique*

Dulong, R., 1978, *Les Régions, l'Etat et la Société locale*, Paris: PUF

Durand-Dastès, F., 1984, 'La question Où? et l'outillage géographique', *Espaces-Temps*, 26-27-28

Duverger, M., 1951, 1981, *Les Partis politiques*, Paris: Armand Colin.

Frémont, A., Chevalier, J., Hérin, R. and Renard, J., 1984, *Géographie sociale*, Paris: Masson

Gatrell, A.C., 1983, *Distance and space : A geographical perspective*, Oxford: Clarendon Press

Gaxie, D., 1978, *Le Cens caché*, Paris: Seuil

Godelier, M., 1984, *L'idéel et le Matériel*, Paris: Fayard

Goguel, F., 1970, *Géographie des élections françaises sous la Troisième et la Quatrième République*, Paris: Presses de la F.N.S.P

Grosser, A., 1972, *L'explication politique*, Paris: Presses de la FNSP

Groupe Dupont, 1982, *A propos de géographie électorale*, Avignon: Brouillons Dupont, Université d'Avignon

Guérin, J.-P., 1984, *L'Aménagement de la montagne : politiques, discours et production d'espaces*, Gap: Ophrys

Guillorel, H., 1983, 'Espace et politique', *Conceptions de l'espace*, Nanterre: Université de Paris-X

Guillorel, H., 1984, 'La géographie électorale des géographes', Paper given at the 2nd National Congres of the AFSP, Grenoble

Johnston, R.-J., 1985, *The geography of English politics*, London: Croom Helm

Katz, E., 1957, *The two-step flow of communication: an up-to-date report on an hypothesis*, French translation. Bourdieu, P. and al.,1982, *le Métier de sociologue*, Paris-La Haye: Mouton

Leleu, C., 1971, *Géographie des élections françaises depuis 1936*, Paris: PUF

Lévy, J., 1981, 'Pour une problématique. Région et formation économique et sociale', *Espaces Temps*, n°10-11

Lévy, J., 1983, 'Vers le concept géographique de ville', *Villes en parallèle*

Lévy, J., 1984, 'Paris, carte d'identité. Espace géographique et sociologie politique', *Sens et non-sens de l'espace*, Paris: Collectif français de géographie sociale et urbaine

Lévy, J., 1985, 'Des citadins contre la ville. Figures décalées, espaces refusés.', Table ronde internationale *Pratique et perception de l'espace*, Pau: Hegoa

Lévy, J., ed., 1991, *Géographie du politique,* Paris: Presses de la Fondation Nationale des Sciences Politiques

Lévy, J., 1994, *L'espace légitime,* Paris: Presses de la Fondation Nationale des Sciences Politiques

Michelat, G. and Simon, M., 1977, *Classe, religion et comportement politique*, Paris: Presse de la FNSP, Editions sociales

Michelat, G. and Simon, M., 1985, 'Déterminations socio-économiques, organisations s ymboliques et comportement électoral', *Revue française de sociologie*, XXVI

Paddison, R., 1983, *The fragmented state. The political geography of power*, Oxford: Blackwell

Rapetti, D., 1985, *Vote et société dans la région nantaise*, Paris: CNRS

Reynaud, A., 1981, *Société, espace et justice*, Paris: PUF

Rouquié, A., 1982, *l'Etat militaire en Amérique latine*, Paris: Seuil

Taylor, P. J. and Johnston, R. J., 1979, *Geography of elections*, Harmondsworth: Penguin Books

Taylor, P. J., 1984, 'Accumulation, legitimation and the electoral geographies within liberal democracy', in Taylor, P.and House, J., eds., *Political geography: recent advances and future directions*, London: Croom Helm; Totowa, NJ: Barnes and Noble

Vanlaer, J., 1984, *200 millions de voix. Une géographie des familles politiques européennes*, Bruxelles: SRBG, Université libre de Bruxelles

Verdès-Leroux, J., 1981, 'Une institution totale auto-perpétuée : la Parti communiste français', *Actes de la recherche en sciences sociales*, 36-37

Jacques Lévy
20, av. Ledru-Rollin
75012 Paris
France

16. Space and communication
A brief analytical look at the concept of space in social theory

Judith Lazar

The space-time dimension of the social context has not often been the subject of investigation by theorists. Social scientists tend to think of time and space as simple dimensions of the environment of action; what is more, they often take for granted the idea of time as linear and measurable. It was in fact geographers who recently introduced into social thought the notion of the convergence of space-time in order to analyse how social development and technological change affect patterns of social activity (Hägerstrand, 1952, 1967; Janolle, 1969). Their studies were based on the notion that all interaction occurs in space, and takes a certain time; social theory cannot therefore ignore the spatial character of interaction through time.

To offer a better understanding of how these social activities are organised within time and space, Anthony Giddens (1984) put forward the concept of regionalisation, a term he does not restrict to its spatial dimension but which furthermore refers to the way in which social life is divided up throughout space-time. Regionalisation, taken in this sense, «should be understood not merely as localisation in space, but as referring to the zoning of time-space in relation to routinised social practices» (p. 119). To make this clear, Giddens gives the example of a private house. A house is a place where a variety of interactions occur throughout a typical day. In present-day societies, houses are divided up into floors, corridors and rooms, and «the latter are also zoned in a different way in time and space» (p. 173). Daytime zones are separated in a clear-cut, logical manner from night-time zones. Thus areas situated on the ground floor are mainly used during the day, while the bedrooms upstairs are reserved for the night.

Briefly, regionalisation concerns the spatio-temporal organisation of social interaction, from various coordination points, in the daily activities of human agents. A place is not merely a spatial location, it is also a framework of

243

G. B. Benko and U. Strohmayer (eds.), Geography, History and Social Sciences, 243–254.
© 1995 Kluwer Academic Publishers. Printed in the Netherlands.

communication, a web of interaction. Consequently, when one wishes to analyse space, one cannot ignore the actors, who, by their interactions, occupy the territory. Any attempts to treat the subject of space refer to social space, since space, the scene of human activities, is defined essentially by its social character. Human action, from the most ordinary to the most heroic, needs an area in which to express itself. Space in its turn structures the interactions it supports. It gives shape, in a more or less constraining way, to values, behaviour patterns, and habits. On this point, it is interesting to remember that in large towns with high immigration, of which Toronto is a prime example, the ethnic districts are a constellation where the customs and practices of different ethnic groups are transmitted within the framework of the unified ways of life of North American civilisation. This «traditional resistance», as Maffesoli calls it (1973), engenders solidarity, and is due to the strength of spatial memory.

Spatiality is a concept which is expressed in different ways. The town can be considered as one of these ways. However, it would be a mistake to restrict spatiality to this single form. If the town (or city) plays a crucial role in the analysis of space, it is because all space is structured on the basis of the town. «The town, the space which engenders sociality» (Maffesoli, 1979). It is in fact within this framework that most human interaction takes place: it is here that the most sociality is to be found. The spaces we are going to look at, from several viewpoints, are always linked round this basic sociality, that of the town. Our aim here is to show, through various social theories, how one can conceptualise space in relation to time and its social character.

1 Social diffusion or innovation in Hägerstrand

The works of the Swedish geographer Torsten Hägerstrand, notably on spatial diffusion, are of particular interest not only to geography, but also in the study of communication. Hägerstrand, in the early 1950s, devoted himself to the problem of the diffusion of knowledge, and more specifically, to the diffusion of innovation in rural Sweden. He focused his attention on the diffusion of «manufactured» innovation, distribution of fertilizers or tractors, among farmers - as opposed to the spread of ideas or behaviour patterns. The basis for his model of innovation is the idea that the adoption of an innovation depends on learning and communication processes. This means that factors relating to the flow of information are the most decisive; consequently, the crucial stage in the analysis of the diffusion process is the identification of the spatial characteristics of the information flow, and the resistance to this adaptation (Brown, 1981).

According to Hägerstrand's theory, information, in the form of a message, stems either from the mass media or from subjects who have already adopted the invention. In a communications network, individuals are at one and the same time receivers and/or transmitter. The transmitters are those who have already adopted the change; the others are the potential receivers. Particular attention has been given to the mechanism of interpersonal communication. Hägerstrand based his conception on the results of empirical work already carried out by American sociologists in the field of communication (Katz, Lazarsfeld, 1948, 1955)[1].

The potential receiver choses to adopt or to refuse the information he perceives. All the same, Hägerstrand notes that the effectiveness of personal messages depends on the interpersonal link established between transmitter and receiver, and he also points out the existence of various barriers - of a social and geographical nature - which may prevent or divert the communication. Geographical barriers, such as lakes, forests, terrains difficult to negotiate, or where the distance between the tow particular speakers is too great, are carefully emphasized.

Hägerstrand supposes that the process of transmission of information takes place in two stages :

- the dissemination of information
- the adoption of the innovation.

The barriers - called «resistance barriers» by Hägerstrand - differ according to the speaker's personal characteristics. The greater the resistance, the more information is needed for adoption to occur as a result. The first hurdle is thus the distance to be covered.

Next, Hägerstrand draws attention to values which are incompatible with the adoption of innovation (social resistance) or to economic conditions which slow down or prevent acceptance (economic resistance). Diffusion on different geographical scales poses the problem of the hierarchy of networks. Thus one network is effective at a local level, another at a regional level. For example, an inter-individual network can be very effective at a local level, notably among farmers (the example tested by the author himself) while for diffusion at a regional level, one can use a central network, made up of people situated at various levels communicating among one another. Hägerstrand himself speaks of hierarchical diffusion when the transmission of information takes place on an increasingly wide scale. The probability of adoption is highest close to a focus of adoption, and decreases with increasing distance. Furthermore, the closer the spatial, familial, socio-economic, ethnic and religious links, the higher the probability of adoption.

Hägerstrand's theory therefore envisages a transformation of the population, starting with a few pioneers, using information through interpersonal communication or media networks. To test his theory, Hägerstrand has carefully worked out maps showing the progression of change in Scandinavia. In a spatial context, like the one he studied, the principal mechanisms of any transformation are the social (interpersonal) communication networks characterised by the distortions (imbalances) and the angles (aspects) characteristic of spatial patterns in the diffusion of innovation. Hägerstrand's diffusion model has had an extraordinary impact on the social science theory and has given rise to innumerable studies.

---

[1] This work has proved, notably, that interpersonal communication is more portant in the changing of attitudes than media communication.

2 Goffman's concept of «social role»

The works of the Canadian sociologist, Goffman, are of interest to us here more because of their relation to social space than to that of physical space. Goffman offers us a panoply of very shrewd and keen observations of encounters, facial expressions, body positionnings, gestures and more generally, of the reflexive control of body movements during the act of communication. The fact that all interaction is furthermore situated, takes place, in space-time, the repetitive nature of everyday life, and the routinisation of activities - a subject particularly emphasized by Goffman - make it possible to draw some connections with our object.

Interactions can be envisaged as the regular and routine occurrence of encounters which take place in time and space. They follow a certain regularity and reflect the institutionalised characteristics of social systems. Social systems exist only in and through the continuity of social practices, which are maintained though encounters that are spread throughout space-time. For Goffman social interactions constitute the very basis of social order, as they are founded on rules and norms. They appear banal and routine, both to the social actors who «play» them, and to the observer trying to analyse them, to pick them out of the flood of ordinary daily actions. However, it is precisely during the most ordinary, most routine encounters, that revealing social issues can be made apparent to the observer.

Goffman pays special attention to the study of encounters, which he considers to be the vital thread of social interaction. Studying the rituals of everyday social life, he came to note that the «positioning» of the body is of fundamental importance in space-time encounters. Encounters always bring into play spatialization, body positioning, and facial expression. He observes that each person takes on multiple stances in space, and hence in social relations which depend on precise social identities. This positioning presupposes numerous types of body movements and gestures. He paid close attention to movements of the face, which he considered to be the dominant area of the body for the human being. The face influences, in a barely perceptible way, the spatial disposition of the people who are interacting. Face to face interaction, for example, necessitates a very particular positioning of the body. In most societies, the act of turning one's back on the person speaking to you during an encounter, is considered to be a highly impolite gesture. Face working, and the monitoring of the body in physiognomic work, is a fundamental dimension of social integration in time and space. In the course of daily life, the actors meet in the context of specific interactions where they are physically in one another's presence. The social characteristics of the encounter are anchored in the spatial disposition of the body. The encounters can be fortuitoius, brief - for instance, whenever actors merely glance at one another - the raising of an eyebrow, a simple gesture, etc., or, on the contrary, take place in more formal contexts, clearly defined in time and space. In the latter, it is often the case of an encounter involving several people at a social gathering. As Goffman puts it, a social occasion provides the «structuring social context» in which several groupings are likely to take place, to split up, and to reform (1963: 18).

Various social occasions can take place simultaneously in physical space, but more often, only one occurs within a given sector of space-time. Within these defined spaces, the actors act out a scene, often in accordance with norms. The concept of «norm» plays a fundamental role in Goffman's theory. He sees it as s sort of «guide to actions sustained by social sanctions» (1973: 101). Throughout all the social interactions of daily life, human beings fill the roles appropriate to social norms. The scripts are written, the stage is set, and the actors are there to play their part. Goffman's conception of social role relies on concepts developed by Linton (1936), who associates social position with role. Goffman, while accepting this proposal, develops the «theatrical» dimension of the concept. Thus the web of daily life is spun with situations presented by actors who play the parts according to routine. Daily life is made up of routine elements. These routines are firmly entrenched in habits, traditions, yet, contrary to what one may think, they are not carried out automatically, without thought on the part of the actor. Thus the control of facial expression (facework), the way the body is held in a specific setting, are the result of a process of reflection. To sustain a routine activity needs the continuity of social practices. In the case of social change, the thread of routine breaks, and new lines of routine conduct are formed within new frameworks. For Goffman, the «presentations» occur in specific settings. The settings contribute to the constitution and the regulation of the activities by defining them. The framework both helps the presentation and at the same time, defines it. Goffman's proposals thus show clearly and eloquently that the physical space surrounding individuals is of a highly social nature.

3 Personal space or the cultural dimension of space in Hall's work

Even though Hall was not the first theorist to take an interest in the cultural dimension of space[2], he is the researcher who most thoroughly analysed the «language» of space. This American anthropologist was preoccupied with the «hidden dimension» of culture, with what is meant by the relationship of «being with space». He believes that each culture organises space in a different way on the basis of its own «territory». Examining animals, for instance, Hall was struck by the crucial importance of territoriality as a behaviour system, «a system which developed in a very similar way to that of anatomical systems». (1966: 23)

Hall studied the perception of space throughout various eras in different cultures, and he discovered that the human being's perception of space is dynamic because it is linked to action (to that which may be accomplished in a given space). He proposes a scale of interpersonal distances, drawn up from

---

2 F. Boas and Sapir, were just two of the more famous precursors.

observations and interviews carried out with a group of American middle class men and women. These personal spaces are divided into two : near and far.

The distance chosen depends on the interindividual relationships, on the feelings and the activities of the individuals. Hall stresses that each human culture has defined in a different way the dimension of the following four zones .

i) *Intimate distance* is reserved for a tiny proportion of social contacts, during which the distance varies between touch (physical contact) and 40 cm. During these contacts the presence of the other is imposed in a voluntary way - a sexual relationship or a parent-child relationship - or it is experienced in a disagreeable way - physical attack, aggression etc. - where the players are submitted to physical contact with the other. The practice of intimate distance is not permitted by adults in American society, even though in certain cases, in public transport, a full lift, a queue etc., they have to put up with it. However, in these cases they have developed a sort of «defence system», such as the contraction of the muscles of that part of the body which comes in to contact with the Other.

ii) *Personal distance* indicates the fixed distance between individuals who have no physical contact. Hall sees it as a sort of bubble which surrounds human agents in order to isolate them from each other. His close modality varies between 45 and 75 cm. It characterizes the majority of intimate encounters between individuals. His distant modality goes from 75 to 125 cm, the distance from easy contact to the point where actors would be hard put do touch each other (by stretching out their arms). It is a case therefore of the «limit of one's physical hold over others». Over and above this limit, it becomes difficult to touch the other person. At this distance the actors have a very clear and precise view of the other, they can distinguish facial peculiarities, movements, etc. This distance is often used in American society between people who know each other fairly well : friendly relations.

iii) *Social distance* characterizes the distance between the personal and the remote world; it goes from 120 cm to 360 cm. The near mode (120 to 210 cm) characterizes the majority of interpersonal relationships in American society, notably between individuals who work together. However, in the more formal professional relationships, it is the distant mode which comes into play. In particular, in the office of an important person, the dimensions of the desk keep visitors at a considerable distance. At this distance the agents cannot distinguish the more subtle facial details, etc. of the person they are speaking to, but they are still able to see clearly the hands, the nails, the condition of the hair, the teeth, etc.

iv) *Public distance* is the fourth category of distance, the largest one, going from 360 cm to 750 cm and even further. Over and above 360 cm (the distant mode) detailed perception of an individual becomes blurred. This distance is characteristic of public performance : words spoken to a public gathering for example. At this distance non-verbal communication, the expression of gestures and body posture become more important, as the human voice has more and more difficulty in conveying nuances of expression. The rhythm of elocution slows down, and the speaker must watch his or her articulation.

These four distances make up four different territories which characterise, for Hall, every type of human society. However, these distances vary according to the culture. Let us remember here that the distances quoted above were drawn up according to North-American culture; they are very different for example in Middle-Eastern culture. It is interesting to remember that Hall did not define the given distances uniquely in centimetres, but also according to other perceptions; sight, hearing, and smell all help in setting limits to socially appropriate distances.

4 The notion of public space in Sennett and Habermas

R. Sennett and J. Habermas are the two present-day theorists who have most closely studied public space. Sennett's approach (Sennett, 1974) is focused on the genesis of public man. He analyses the conditions of his birth, of his existence and the reasons for the «fall» of public space, within which public man flourished. According to Sennett, public space was created historically at a time when it was still possible to communicate anonymously, without individuals being obliged to reveal their personal identity in order to give some credibility to their words. This anonymity therefore implied an egalitarian dimension, as the social differences were totally unimportant in the action of communication during which these distinctions were left aside. What was said was then validated by the force of argument. With the growth of a hierarchical society, and the emergence of industrial capitalism, this type of exchange was doomed to disappear. Sennett devotes a large part of his description to large towns, such as Paris and London, which have been particularly favourable to social exchange, and to the presentation of social life. Squares in large towns are places where strangers pass each Other, and meet daily. In this context public man flourished, communicating anonymously, and ready to accept the impersonality of social exchange.

The stable balance which existed throughout the eighteenth century between public and private space - reserved for family life and friends - was upset at end of the century by the birth of industrial capitalism. The end of public space did not happen overnight, it continued through the 19th century. The emergence of industrial capitalism highlighted the private sphere to the detriment of the public one. During the 19th century, the family became one of the bastions of protection, in a hostile outside world where the economic and political struggle was more and more ferocious. Thus, private space became a refuge with a moral value considerably higher than public space, considered as morally degenerate. Therefore, while public space was depleted, private space gained in importance. In the 19th century, public space became the «dangerous» place where a well-educated woman did not venture alone.

In this increasingly intimate society, public behaviour has changed overall. The streets are now silent, exchanges reduced, and the individuals themselves are less and less expressive. We are far from the image of a society which has been blithely compared to a theatrical stage. The theatrical side of life disappears in favour of a general transparency. While public man in the *Ancien régime* was gifted with a talent for investing in the social game, nineteenth century man refuses the game, or the rite. For Sennett, public space, the site of social exchange, can no longer function in a world which refuses the game, which refuses anonymity, and is only concerned with social position. If the author draws a parallel between a theatre stage and eighteenth-century «theatricality», it is because he recognises in both the same expression used as a means of communication. There is in fact a certain affinity between them. Both take place in a world where the public is made up of heterogeneous individuals, ready to accept the rule of presentation. As public life fell to pieces, so this affinity crumbled. When public life is strong, social expression is seen as presentation, whereas it becomes representation with the coming of industrial society. When private space mingles with public space, the border between the two disappears and even sociality diminishes.

All in all, the «end» of public man, in the eyes of Sennett, is a direct result of the end of the theatrical game, the end of conventions, and in the end, of the routinisation of symbolic mediations, on which up till then the social bond had relied. The emergence of the «new man», nineteenth-century man, the bourgeois, the product of industrial capitalism, is closely linked to this decline. The codes have changed; the normative model has become the visibility of one's opinions, of feelings. Thus, if the public space has become empty, it is because communication techniques have taken over from theatrical role-plays. This changeover was not entirely painless : social links suffered, just as on an individual level the social actors, in losing interest in the game, have lost, as it were, their identity.

Habermas, unlike Sennett, is interested less in the presentation of the self than in the directly communicative field in which the action takes place (Habermas, 1989). Through a socio-historical analysis of the complex of meanings which revolve around the term «public», he tries to recreate the historical conditions which contributed to the birth and the structuring of this space. He neatly demonstrates that the conditions which more than anything contributed to the formation of a communication field are those of the social differentiation which took place in the structures of society at the time of the transition from feudalism to capitalism.

In the Greek town, public life was lived in the market place, the agora, though it was not dependent on this site. The public sphere exists within dialogue. At this period, public space is totally separate from private space, which belongs to each individual. During the European Middle Ages, these differences continue to exist, even though they are less restrictive. The public sphere of the Middle Ages is structured by the representation of power, like the prince's seal or his badges. The Reformation put an end to this representation. The position of the Church underwent a decisive change; the site of religious authority which it represented became a private matter. This phenomenon was followed by the break-away of other institutions from royal power. Habermas stresses the very clear separation between the public sphere - royal power - and the private sphere, based on the full use of property of the capitalist type.

The development of the nuclear families model was also a decisive evolutionary factor. This interior space is the scene of an emancipation at once psychological, economic and political. Once the reproduction of the means of survival depends on generalised social exchanges between property-owners, the role of the family undergoes an important modification. The bourgeois family becomes the site of the production of a new awareness, combining independence, affection and culture. It is within the framework of the bourgeois family-unit that the sense of individuality develops which fuels the need for cultural consumption. Habermas insists that liberal public sphere was initially formed within the field of culture. In parallel with the Court at the end of the seventeenth century, a cultural or literary public sphere can be seen to emerge which soon comes to oppose the Court. Habermas' analysis attributes a very important role to this cultural space based on specific institutions such as cafes, salons, and clubs, as it was there, according to him, that the principles of the rules and the structures of public space were determined and finalised, before spreading to the political sphere. These sites, in spite of their different type of clients, had no social hierarchy. It was there, in a place which escaped control by state power, that Reason could blossom unhindered. From the beginning of the 18th century, cafes became so numerous, their circle of clients so wide, that it became necessary to create newspapers for them. «The Tatler», «the Spectator», and the «Guardian» became veritable institutions and offered subjects of discussion to a wide public of bourgeois origins. Thus the press became a public organ of the bourgeois strata, personifying the public consciousness which corresponded to a literary use of reason. The emergence of this cultural space where «criticism is used against the power of the State», is accomplished like a subversion of the public literary consciousness, already endowed with a public possessing its own institutions and platforms for discussion (p. 61).

The public sphere should therefore be seen, according to Habermas, as the place where a society founded on rational communication was organised. It is thanks to the advancement of Reason that it was possible to free art and culture from subordination to political power institutions. Literary and philosophical works, and works of art, were produced for and distributed by a public market, and in principle therefore became accessible to everyone. However, this «general public» nevertheless remained limited, made up mostly of enlightened, wealthy Bourgeois. Little by little, a professional «higher» stratum detached itself from the public, and progressively took over a position of authority as judges of the arts, defining themselves as the spokesmen of the general public, and filling the double role of representative and educator. Their aim was to enlighten minds in order to free them. A political awareness gradually grew within this bourgeois public sphere which demanded the rationalisation of abstract laws and a new source of legitimacy.

The public sphere thus created served basically to defend the private domain of the economy. The aim was to determine and to control the new rules of exchange, and to limit the intervention of authority in the management of private interests. This public opinion therefore became a sort of mediator between the State and newly developing needs. The birth of the public and political sphere is closely linked to a wish for and an obligation for complete emancipation. Because to gain access to this public space, to defend his legal rights, the individual has to be master of him-self.

A decline, however, was not long in coming. Habermas sees the reason for this decline as tied up with the main criteria of emancipation itself : culture and property. The public sphere is founded on the basis of private interests, based on the freedom of enjoyment of ownership. Contradictions soon make their appearance. The public domain gradually loses its role of unifying the public interest and ceases to be the site of a critical discussion. The media, by which the system was supported in the beginning, became instruments of manipulation, agents of integration into State norms, in the hands of powerful social groups. The public sphere became progressively depoliticised, its critical function wearing away, an came to disappear totally towards the middle of the 19th century.

5 Innis and the dialectical claim for space and communication

If this strange journey through space ends with Harold Innis, it is not in order to do him justice, even though there is no doubt he deserves to have a great deal of thought given to his work, which is extremely rich but little or badly known. Nor is it to find reassurance, a faultless clarity after the theories which have been examined and require examination. It is simply because Innis' ideas provide knowledge on a particularly broad scale, which encompasses, as it were, all the dimensions of the analysis of space and of communication.

Neither Hägerstrand, nor Goffman, nor Hall, any more than Sennett or Habermas, are familiar with his work, or else they simply do not speak of it. Nevertheless, the fact remains that we find the germ of almost all their ideas in the theories of Innis. Thus, it does not appear to me to be at all reasonable to talk of space, in what ever way, without appealing to Harold Innis. His most original contribution, and perhaps also the most important, was his analysis of the problematic of technology in terms of a new fusion between time and space. In Innis' account, the historical process may be analysed as a continual struggle between the dimensions of time and of space throughout different civilisations. As each civilisation has a dominant form of communication, the result is an imbalance in cultural orientation with regard to time or to space. An equilibrium between these two factors is relatively rare.

Our knowledge of other civilisations depends for the most part on the characteristics of the dominant media used by each epoch. A communication medium has a decisive influence on the propagation of knowledge in space and in time. This is why, according to Innis, it is necessary to study its characteristics in the right order, so that we can evaluate its influence in a given culture. According to its characteristics, one medium is better suited than another to the spread of knowledge though time or space. Thus heavy and durable media, which are not easy to transport, characterise societies known as «time-biased»; we are then talking of traditional societies, with little spatial extension where the basis of communication has been the clay tablet, stone or parchment. These societies were characterised by the continuity and the survival of their traditions and sacred rites. Traditional societies favoured decentralisation and hierarchical institutions, and also enabled the formation of a powerful class, the only one enjoying access to knowledge. Monopoly of knowledge became a powerful means and a fearful weapon in the regulation of the division of labour of the people. These rigid civilisations were often hard to maintain, because of the various tensions which erupted in their midst.

The introduction of a lighter, more easily transportable medium, such as papyrus or paper, led to a different social structure. These media were not as long-lasting, but thanks to their lightness, were particularly suited to the administration of large territories and to trade. The Roman conquest of Egypt gave access to supplies of papyrus, which became the basis of the administration of the Roman Empire. In Southern Europe, paper and later the printing industry, favoured the national languages, and posed a challenge to Latin, which was associated with the stiffer parchment. These lightweight supports favoured centralisation, and a less hierarchical system of government, less concerned by tradition and looking more to the present and the future. These civilisations are characterised by a complex political authority and by the creation of abstract science and technological knowledge.

Innis also devoted a considerable part of his explanation of the birth and decline of a civilisation to the environment. The environment appeared to him not as an unchanging framework, more as an active agent in the territorial construction of a country. He does not consider the territorial construction as decisive. Actors do not submit to it passively, on the contrary, they can challenge it. For Innis, technology was conceived as an obligatory passage for interactions between society and the environment. If it allowed society to change its environment, it is equally true that society also transformed itself in the process. He showed how technologies have allowed the emergence of new territorial organisations or, on the contrary, the preservation of old ones. For him, territorial construction is inseparable from communication and from the communication techniques which the individuals living in the given territory have chosen.

The means of communication are decisive in the social organisation of a given society and they maintain a close link with the powers that be. Innis has always attributed an important role in his analysis to social and political effects. The originality of Innis's approach from a geographical perspective, is that he was able to envisage this object of study in the interaction between the environment and the adoptive values of a society in the two dimensions of time and of space.

Here we have to end our survey, which does not pretend to be exhaustive or to offer an in depth analysis of the idea of space. Our main concern has been to demonstrate the relationship, sometimes barely perceptible, which links every communicative action to space and to show that this is one of the most important factors to be taken into consideration.

References

Brown, L. A., 1981, *Innovation diffusion. A new perspective*, London: Methuen

Giddens, A., 1984, *The constitution of society*, Cambridge: Polity Press

Goffman, E., 1963, *Behaviour in public places*, New York: Free Press

Goffman, E., 1969, *The presentation of self in everyday life*, Harmondsworth: Penguin

Habermas, J., 1989, *The structural transformation of the public sphere: An inquiry into a category of bourgeois society*, Cambridge, MA: MIT Press (Original work publ. 1962)

Hägerstrand, T., 1952, *The propagation of innovation waves*, Lund: Gleerup, (Lund studies in Geography)

Hägerstrand, T., 1967, *Innovation diffusion as a spatial process*, Chicago: University of Chicago Press

Hall, E. T., 1966, *The hidden dimension*, New York: Doubleday

Innis, H., 1950, *Empire and communication*, Oxford: Oxford University Press

Innis, H., 1951, *The bias of communication*, Toronto: University of Toronto Press

Janelle, D. G., 1969, 'Spatial reorganisation: a model and concept', *Annals of the Association of American Geographers*

Katz, E. and Lazarsfeld, F., 1955, *Personal influence: the part played by people in the flow of mass communications*, Glencoe, IL: Free Press

Lazar, J., 1992, 'La compétence des acteurs dans la théorie de structuration de Giddens', *Cahiers Internationaux de Sociologie*, 39, 399-416

Linton, R., 1936, *The study of man*, New York: Appleton Century

Maffesoli, M., 1979, *La conquête du présent*, Paris: PUF

Quéré, L., 1982, *Les miroirs équivoques, Aux origines de la communication moderne*, Paris: Aubier-Montaigne

Sennett, R., 1974, *The fall of public man*, New York: Vintage Books

Judith Lazar
Fondation Avicenne
27 d, bld. Jourdan
75690 Paris Cedex 14
France

PART VI

CONCLUSION

17. Conclusion : The spatialization of the social sciences

Ulf Strohmayer

Few are those to deny that the current state of the social sciences is one of crisis. Brought about by a widespread recognition of the instability inherent to social scientific concepts, the very idea of a *crisis* appears to have turned ubiquitous - so ubiquitous indeed as to constitute a new and commonly accepted status quo. In the wake of postmodernism, going round in circles has become the norm and an emerging consensus within the social sciences holds that where everything has become questionable in general, we no longer need to question anything in particular.

Traditional critical theorists, of course, were among the first to lament the stasis of the current crisis. Their interests in legitimizing the particularities of critical interferences in general terms were ill served by the factuality of a non-discriminating and non-teleological «postmodern» crisis. Problematic, as many of the offered solutions to this impasse remain in their implicit or explicit appeal to a pragmatic use of reason, even the less strategically inclined participant in the current debates will by now feel exasperated with the protection of business as usual everywhere within the social sciences.

Chief amongst the consequently unquestioned understandings is the common belief that some kind of conceptual rationality lies dormant behind the divisions dividing the social sciences into various disciplines. In assembling the present volume, Georges Benko and I felt it necessary to remind the reader that there is nothing natural about any of the existing disciplinary divides. On the contrary - on many occasions do they serve as counterproductive barriers to insights rather than help to illuminate the social world. We launched this reminder from *within* geography - not in order to develop yet another confusion of the «inside» «outside» kind but because we are both geographers by training and had thus ample of opportunities to appreciate geography's integrative approach to social realities. To be sure, this integrative quality inherent to most of the work done in geography quite often brings with it a certain dilettantism,

257

G. B. Benko and U. Strohmayer (eds.), Geography, History and Social Sciences, 257–260.
© 1995 *Kluwer Academic Publishers. Printed in the Netherlands.*

especially if its depths are evaluated from the vantage-point of more specialized disciplines. Yet today, with the legitimacy of questioning claims to conceptual expertise no longer denied, this very dilettantism appears more as a *possibility* to generate insights than as the roadblock on the path of an ever-more refined knowledge we used to associate with «dilettantism» in the past. A plea for the necessity of geography is thus decidedly *not* an argument in favour of one particular discipline but instead advocates a mode of approaching «world» in general.

This manner of thinking called «geography» seems all the more necessary today since the above mentioned postmodern breakdown in the legitimacy of stable concepts brought with it the recognition of a need to «spatialize» epistemology. Walking in the footsteps of such diverse «spatializers» as Henri Lefebvre and Anthony Giddens and working both with and against colleagues in geography such as Allan Pred, Gunnar Olsson and Michael Dear, it was Ed Soja and his *Postmodern Geographies* (1989) in particular who made us see how the traditional epistemological priorization of time over space has been detrimental to any of the resulting «knowledges». Linearized and immanently teleological, they became bereft of the motor for change - breaks and fissures within the smooth unfolding of events, i.e. social tensions that spring forth from an *always already* uneven materialization or «taking place» of these events. In fact, events never are what they are but through the fact that they are what they are *somewhere*. This «somewhere, rather than nowhere» consequently became the epistemological equivalent to Martin Heidegger's ontological «something, rather than nothing»; in both cases the problem proved to be one of linkage and otherness. Just as the ontological recognition of «being» creates its *other* in «nothingness», the epistemological recognition of generalities yields its own particular *other* in geographic specificity. What is more, *specific otherness* in general is in turn always already (if implicit in most cases) constitutive of general concepts and thus detachable from generalities only at the risk of altering them beyond recognition. A «spatialized» epistemology, in other words, would fall back into a provincial and revisionist priorization of geography over the ensemble of other social sciences if it were to neglect or, worse, to negate this abstract realization. Time and space, general and particular are but two sides of the same coin and any analytical separation of the two can only falsify a resulting understanding of social reality.

We may add to this abstract train of thoughts the growing awareness among social scientists that the prolonged reliance on a-spatial concepts or the scientific preference for justification over the «mere» possibility of insights has been a mixed blessing at best. Gains in general knowledge all too often failed both to answer pragmatic needs and to encompass, rather than exclude, vital differences of scale. The currently widespread loss in confidence especially in dualistically constructed concepts like, for instance, «man» and «nature» or «urban» and «rural» originated at least partially in the recognition that these concepts, *if taken statically*, generate as many problems as they might help to overcome. It is within this practical context that the articles presented in this volume mark one step towards a «spatialized» social science. Within its own topological focus, each of these articles approaches a highly unique (and thus certainly debatable) reconciliation between necessarily general concepts and unavoidably particular otherness. In refusing to think of these reconciliations in terms of new generalities, the articles at hand become exemplary attempts at re-thinking the current crisis without betraying its motifs. They invite us to reconsider our individual sets of stable concepts by presenting historical, geographical or epistemological differences - the ubiquitous possibility that, if seen through a differently conceptualized set of lenses, our understanding of the social world might change tremendously. The concepts of «man» and «nature», for instance, change both their respective shapes and contents in the articles by Santos and Staszak; «urban» and «rural» lose the sense of natural «givenness» often attributed to this pair in the papers by Hägerstrand, Mitchell-Weaver and myself; and the overarching question of the relationship between generality and specificity is conceptualized anew (and, admittedly, in strikingly different forms of expression) in the contributions by Gregory, Hampl, Claval and Johnston. Here, as in all of the articles in this collection, an oft-invoked *spatial difference* represents more than a mere «relativization» of otherwise untouched and non-modified general lines of argumentation; instead, it becomes an integral component of any effort to understand the social world.

Most readers will associate these and related efforts to redirect the social sciences towards «spatial» epistemologies with postmodernism in general. Personally, I hesitate to concur with this indeed common interpretation. While it is certainly true that the impulse to acknowledge a fundamental arbitrariness at the root of conceptual distinctions owes much to the emergence and, of late, the respectability of postmodern discourses, far too many continuities between «spatialized» epistemologies and traditional pursuits of knowledge in the social sciences remain to speak of a clear-cut break or «post». To start with, *any* epistemology has as its root the differentiation between «knowledge» on the one hand and an-Other, some non-reflective lack of knowledge on the other. In this respect at least, none of the articles presented here or any of the attempts associated with «spatialization» in general does break with the modern idea of an *a priori* differentiating status which baptizes emerging insights. Also, and closely related to this first point, it is hard to avoid the conclusion that the quintessentially modern idea of *progress* is altogether absent from much of what passes as «spatialized» social science. Furthermore, the rhetorical devices employed here and wherever space is «reasserted» do not differ dramatically from those of older epistemologies; respectability still is awarded to those who accept a predetermined style of argumentation as binding. With the exception of Allan Pred's work, «spatial» epistemologies still express themselves a-spatially.

None of the above, allow me to stress, should be mistaken as a sign of criticism of particular efforts - it is after all far from settled whether or not one can avoid *being read* as embodying knowledge, as being progressive or a-spatial - but as skepticism in view of a common characterization of these efforts as «postmodern». In fact, as I have argued elsewhere, this skepticism extends to the usefulness of the very term «postmodern» in general (1993). Yet postmodern or not, what we can witness today in geography and elsewhere in the social sciences clearly embodies the promise of rejuvenation. We should accept, though, that a rejuvenated social science will not make life easier for any one of us engaged in any part of it. Rather, we will encounter questions where before we were lead to expect answers. We may even learn - and here a trans-national collection of essays like the present one could certainly be of help - that the manners according to which we learnt to phrase questions are themselves not at all automatically compatible with one another. There are many facets to this project we call «Enlightenment» and the more of these we encounter the better for whatever it may turn out to have been that we now seek to enlighten. This may seem a strangely idealist position at first and one which furthermore makes it difficult to derive a short-term benefit from the social sciences, but have we not been chasing «the short term» too long already ? Above all, Enlightenment is a process which necessarily and continuously involves the labour of understanding. Especially in times of change such as today we may be well advised to *hesitate* before rushing to conclusions.

References

Benko, G. and Strohmayer, U., eds., 1996, *Space and Social Theory: Geographic Interpretations of Postmodernity,* Oxford: Blackwell, (forthcoming)

Soja, E., 1989, *Postmodern Geographies. The Reassertion of Space in Social Theory*, London: Verso

Strohmayer, U., 1993, 'Modernité, Postmodernité, ou, Comment Justifier un Savoir Géographique?' *Géographie et Cultures*, 6, 75-84

List of figures :

CONTRIBUTORS

BENKO Georges
Professor of geography at the University of Panthéon-Sorbonne. He currently works at CRIA (Paris). He is a member of the editorial board of *Espaces et Sociétés* and *GeoJournal* and editor of the series *Géographies en Liberté* and *Théorie Sociale Contemporaine* (published by l'Harmattan). He is the author of *Géographie des technopôles* (Masson) and editor of *Les nouveaux aspects de la théorie sociale* (Paradigme), *La dynamique spatiale de l'économie contemporaine*, (Editions de l'Espace Européen) and *Industrial Change and Regional Development* (with Mick Dunford, Belhaven Press/Pinter), and co-wrote together with Alain Lipietz *Les régions qui gagnent. Districts et réseau: les nouveaux paradigmes de la géographie économique* (P.U.F.)
Address: Université de Paris I - Panthéon-Sorbonne, 191, rue Saint-Jacques, 75005 Paris, France

CLAVAL Paul
Claval is one of Europe's foremost geographers and professor at the University of Paris-Sorbonne. He is a founding member of the editorial board of *Géographie et Cultures,* author and co-author of about 30 books, including *Les mythes fondateurs des sciences sociales*, (1980), *Géographie humaine et économique* (1984), *Geography Since the Second Word War* (1984), *La conquête de l'espace Nord-Américain* (1989), *La géographie au temps de la chute des murs* (1993), *Géographie de la France* (1993), *Initiation à la géographie régionale* (1993), *Géopolitque et géostratégie* (1994).
Address: Université de Paris IV - Sorbonne, 'Espace et Culture', 191, rue St.-Jacques, 75005 Paris, France

DUNFORD Mick F.
Mick Dunford is professor in human geography in the School of European Studies, University of Sussex. He is a member of the editorial board of *Espaces et Sociétés*. His main publications include *The arena of capital* (Macmillan, 1983) written with Diane Perrons, and *Capital, the state and regional development* (Pion, 1988). He is also the editor (with G. Benko) of *Industrial change and regional development: the transformation of new industrial spaces* (Pinter/Belhaven Press, 1991) and *Cities and Regions in New Europe* (with G. Kafkalas, Pinter/Belhaven Press, 1992). At present he is doing comparative research on the development of new technologies in Europe and is working on the regional implications of European economic and political integration.
Address: School of European Studies, University of Sussex, Arts Building, Falmer, Brighton, BN1 9QN, United Kingdom

FOUCAULT Michel (1926 - 1984)
After having occupied a host of different positions at different institutions in France and abroad, the late Foucault held positions at the University of Paris VIII (Vincennes) and from 1970 until his untimely death in 1984 occupied the chair for the History of Ideas at the Collège de France in Paris. Most of his work has been translated into a large number of foreign languages including *Maladie mentale et psychologie* (1953), *Histoire de la folie à l' âge classique* (1961, 1972), *Naissance de la clinique* (1963), *Les Mots et les Choses* (1966), *L'Archeologie du savoir* (1969), *L'Ordre du discours* (1972), *Surveiller et punir* (1975) and the three volumes of his *Histoire de la sexualité* (1976-1984).

GREGORY Derek
Born and educated in England, Gregory is professor of geography at the University of British Columbia (Vancouver). He has taught previously at the University of Cambridge (G.B.). Gregory is co-editor of *Society and Space*. He is the author of various publications on geography. His publications include: *Ideology, Science and Human Geography* (1978), *Regional Transformation and Industrial Revolution* (1982) *Social Relations and Spatial Structures* (with J. Urry eds., 1985), *Horizons in Human Geography* (with R. Walford eds., 1989), *Geographical Imagination* (1994). He is currently exploring the connections between information circulation, the production of 'imagined geographies' and the impress of imperial power within the British Empire in the ninetheenth and early twentieth centuries.
Address: Department of Geography, University of British Columbia, 1984 West Hall, Vancouver V6T 1Z2, Canada

HÄGERSTRAND Torsten
Professor of Human Geography at the University of Lund, Sweden, Hägerstrand is most famous for his development of «time-geography». Hägerstrand has been one of the leading voices within the geographic community for more than three decades. His most important book is *Innovation Diffusion as a Spatial Process* (1968); Professor Hägerstrand received the International Geographical Price (the 'Nobel of geographers') in Saint-Dié-des-Vosges in 1992.
Address: Box 716, 22007 Lund, Sweden

HAMPL Martin
Professor Hampl is the head of the Department of Social Geography and Regional Development at Charles University in Prague; his interests include, among others, theoretical geography, territorial administration and urban geography; he is currently the Czech coordiator of the joint research project on «Regional and Local Development» with Amsterdam University, the Netherlands. He has published numerous books and articles in Czech and in English.
Address: Department of Social Geography and Regional Development, Charles University, Albertov 6, 12843 Prague, Czech Republic

JOHNSTON Ron J.
Formerly professor of geography at Sheffield University, Johnston is currently Vice Chancellor of the University of Essex at Colchester. He received the Murchison Award from the Royal Geographical Society. He is series editor at Belhaven Press/Pinter Publishers. Author and editor of numerous books including, *The Future of Geography* (1985), *Philosophy and Human Geography* (2nd ed., 1986), *On Human Geography* (1986), *Environmental Problems: Nature, Society and the State* (1989), *Developments in Electoral Geography,* (1990), *Geography and Geographers* (4th edition 1991), *A Question of Place: Exploring the Practice of Human Geography* (1991), *The Changing Geography of the UK* (1991), and co-editor of *The Dictionary of Human Geography* (3rd edition, 1994)
Address: University of Essex, Wivenhoe Park, Clochester CO4 3SQ, United Kingdom

LAZAR Judith
Ph.D in Sociology, Judith Lazar is a professor in sociology at the University of Paris X - Nanterre. She is the author on many books including: *Ecole, Communication Télévision* (Paris, PUF, 1985), *La télévision: mode d'emploi pour l'école* (Paris, ESF, 1988), *Sociologie de la communication de masse* (Paris, A. Colin, 1991), *La science de communication* (Paris, PUF, 1992), *L'opinion publique* (Paris, Dalloz/Sirey, 1995)
Address: Fondation Avicenne, 27 d, bld. Jourdan, 75690 Paris Cedex 14, France.

LÉVY Jacques
Student of the École Normale Supérieur at Cachan, Lévy is currently professor at the Institut d'Études Politiques at Paris and at the University of Reims. He is founding member of the journal *EspacesTemps*. He has served on numerous boards and committees, notably within the International Geographical Union. His most recent publications include *Révolutions, fin et suite* (with Patrick Garcia and Marie-Flore Mattei, EspacesTemps/Centre Georges Pompidou), *Le monde: espaces et systèmes* (with Marie-Françoise Durand and Denis Retaillé, Presses de la FNSP/Dalloz, 1992) and *L'espace légitme* (Presses de la FNSP, 1993).
Address: 20, av. Ledru-Rollin, 75012 Paris, France.

PERRONS Diane
Diane Perrons is professor in the Department of Economics at the City of London Polytechnic. Her publications include *The arena of capital,* (MacMillan, 1983), written with M. Dunford. Her current research interests are concerned with the economic and social implications of technical change, in particular the role of telecommunications in regional development. She is also interested in gender issues, in particular equal opportunities policies.
Address: 57, Upper Abbey Road, Brighton, BN2 2AD, United Kingdom

SANTOS   Milton
Professor of Geography at the University of São Paolo, Brasil, and president of ANPEGE. He has taught at a number of universities in Brasil and abroad (Toronto, Paris, New York, Lima, Bordeaux, Toulouse, Caracas, Rio, Bahia). He is the author of numerous publications, including *Les villes du Tiers Monde* (1971), *The Shared Space* (1973), *L'espace partagé* (1975), *Pour une géographie nouvelle* (1984), *Espace et méthode* (1990) *Técnica, Espaço, Tempo: Globalização e meio tecnico-cientifico informacional,* (1994) and has received the International Geographical Price (the 'Nobel of geographers') in Saint-Dié-des-Vosges in 1994.
Address: Rua Nazaré Paulista 163, apt. 84, 05448-000 São Paulo, Brasilia

STASZAK   Jean-François
Professor at the University of Picardy (Amiens), he is a associated with the research center *Espace et Culture* (Unversity of Paris-Sorbonne, CNRS), the editorial board of *Géographie et Cultures*. Author of numerous articles and books: *La goudron dans la brousse. La Route de l'espoir (Mauritanie)* (1989) and *La géographie d'avant la géographie. Le climat chez Aristote et Hippocrate* (1995).
Address: Espace et Culture, Université de Paris-IV, 191, rue Saint-Jacques, 75005 Paris, France.

STROHMAYER   Ulf
After studying at the Universities of Munich, Freiburg, Paris-Sorbonne, the Fernand Braudel Center (SUNY, Binghamton), Nordplan in Stockholm and receiving his doctorate from Penn State University, Strohmayer is currently professor in human geography at the University of Wales, Lampeter. He is the author of numerous articles, co-editor of two books and author, together with Matthew Hannah, of *Gnostic Materialism. Cosamology and the Foundations of Social Theory* (forthcoming).
Address: University of Wales, Department of Geography, Lampeter, Dyfed, SA48 7ED, United Kingdom

WEAVER   Clyde
Professor Weaver is Dean and teaches economic development and planning at the University of Pittsburgh. His books include *Territory and Function: The Evolution of Regional Planning* (with John Friedmann) and *Regional Development and Local Community.* He has been professor at the University of British Columbia, and visiting professor at the University of Paris, the University of Lyon, and the University of Aix-Marseille.
Address: GSPIA, University of Pittsburgh, 3N 22 Forbes Quadrangle, Pittsburgh, PA 15260, USA

INDEX

The GeoJournal Library

1. B. Currey and G. Hugo (eds.): *Famine as Geographical Phenomenon.* 1984
   ISBN 90-277-1762-1

2. S.H.U. Bowie, F.R.S. and I. Thornton (eds.): *Environmental Geochemistry and Health.* Report of the Royal Society's British National Committee for Problems of the Environment. 1985
   ISBN 90-277-1879-2

3. L.A. Kosiński and K.M. Elahi (eds.): *Population Redistribution and Development in South Asia.* 1985
   ISBN 90-277-1938-1

4. Y. Gradus (ed.): *Desert Development.* Man and Technology in Sparselands. 1985
   ISBN 90-277-2043-6

5. F.J. Calzonetti and B.D. Solomon (eds.): *Geographical Dimensions of Energy.* 1985
   ISBN 90-277-2061-4

6. J. Lundqvist, U. Lohm and M. Falkenmark (eds.): *Strategies for River Basin Management.* Environmental Integration of Land and Water in River Basin. 1985
   ISBN 90-277-2111-4

7. A. Rogers and F.J. Willekens (eds.): *Migration and Settlement.* A Multiregional Comparative Study. 1986
   ISBN 90-277-2119-X

8. R. Laulajainen: *Spatial Strategies in Retailing.* 1987   ISBN 90-277-2595-0

9. T.H. Lee, H.R. Linden, D.A. Dreyfus and T. Vasko (eds.): *The Methane Age.* 1988
   ISBN 90-277-2745-7

10. H.J. Walker (ed.): *Artificial Structures and Shorelines.* 1988
    ISBN 90-277-2746-5

11. A. Kellerman: *Time, Space, and Society.* Geographical Societal Perspectives. 1989
    ISBN 0-7923-0123-4

12. P. Fabbri (ed.): *Recreational Uses of Coastal Areas.* A Research Project of the Commission on the Coastal Environment, International Geographical Union. 1990
    ISBN 0-7923-0279-6

13. L.M. Brush, M.G. Wolman and Huang Bing-Wei (eds.): *Taming the Yellow River: Silt and Floods.* Proceedings of a Bilateral Seminar on Problems in the Lower Reaches of the Yellow River, China. 1989   ISBN 0-7923-0416-0

14. J. Stillwell and H.J. Scholten (eds.): *Contemporary Research in Population Geography.* A Comparison of the United Kingdom and the Netherlands. 1990
    ISBN 0-7923-0431-4

15. M.S. Kenzer (ed.): *Applied Geography.* Issues, Questions, and Concerns. 1989
    ISBN 0-7923-0438-1

16. D. Nir: *Region as a Socio-environmental System.* An Introduction to a Systemic Regional Geography. 1990   ISBN 0-7923-0516-7

17. H.J. Scholten and J.C.H. Stillwell (eds.): *Geographical Information Systems for Urban and Regional Planning.* 1990   ISBN 0-7923-0793-3

18. F.M. Brouwer, A.J. Thomas and M.J. Chadwick (eds.): *Land Use Changes in Europe.* Processes of Change, Environmental Transformations and Future Patterns. 1991   ISBN 0-7923-1099-3

The GeoJournal Library

KLUWER ACADEMIC PUBLISHERS – DORDRECHT / BOSTON / LONDON